THE TAIN

THE TAIN

TRANSLATED FROM THE IRISH EPIC
TAIN BO CUAILNGE

BY THOMAS KINSELLA

WITH BRUSH DRAWINGS
BY LOUIS LE BROCQUY

Published in association with
The Dolmen Press, Dublin
by
OXFORD UNIVERSITY PRESS

Oxford University Press, Walton Street, Oxford OX2 6DP

OXFORD LONDON GLASGOW NEW YORK
TORONTO MELBOURNE WELLINGTON CAPE TOWN
IBADAN NAIROBI DAR ES SALAAM LUSAKA ADDIS ABABA
KUALA LUMPUR SINGAPORE JAKARTA HONG KONG TOKYO
DELHI BOMBAY CALCUTTA MADRAS KARACHI

ISBN 0 19 281090 1 Paperback
ISBN 0 19 212545 1 Clothbound

Text © Thomas Kinsella 1969
Illustrations © Louis le Brocquy 1969

First published in a limited edition by
The Dolmen Press, Dublin, 1969

This edition first published, simultaneously in clothbound
form and as an Oxford University Press paperback,
by Oxford University Press, London, 1970
Reprinted 1972, 1974, 1975 and 1977

Published in association with
The Dolmen Press Limited, Dublin

Printed in Great Britain by
Fletcher & Son Ltd, Norwich

CONTENTS

Translator's Note and Acknowledgements *page* vii

Artist's Note viii

Introduction ix

Maps xvii
   I   The Ireland of the Táin
   II The Route of the Táin
   III In Conaille and Cuailnge

Some Recommended Books xxiv

Pronunciation of Irish Words xxv

BEFORE THE TAIN

How the Táin Bó Cuailnge was found again 1

Conchobor:

How Conchobor was Begotten and how he took
   the Kingship of Ulster 3

The Pangs of Ulster 6

Exile of the Sons of Uisliu 8

Cúchulainn:

How Cúchulainn was Begotten 21

Cúchulainn's Courtship of Emer and his Training
   in Arms with Scáthach 25

The Death of Aife's One Son 39

The Quarrel of the Two Pig-keepers and how the
   Bulls were Begotten 46

THE TAIN

   I   The Pillow Talk 52

   II The Táin Bó Cuailnge begins 58

   III The Army Encounters Cúchulainn 65

   IV Cúchulainn's Boyhood Deeds 76

   V   'death death!' 92

   VI From Finnabair Chuailnge to Conaille 100

VII	Single Combat	114
VIII	The Bull is Found. Further Single Combats. Cúchulainn and the Morrígan	125
IX	The Pact is Broken. The Great Carnage	137
X	Combat with Fergus and others	156
XI	Combat of Ferdia and Cúchulainn	168
XII	Ulster Rises from its Pangs	206
XIII	The Companies Advance	224
XIV	The Last Battle	238
	Notes on the Text	255

TRANSLATOR'S NOTE AND ACKNOWLEDGEMENTS

The making of this translation has been very much an aside to other things. It is fifteen years since I was first tempted to do it. I had just found the oldest version of the Deirdre story and been struck by its superiority over the usual one, and I thought I would look closer at the rest of the Ulster stories. I was unprepared for the difficulties in the way of this mild curiosity. There were plenty of 'retellings' in the book-shops, but actual translations were scarce, and those I could find were generally dull. I emerged with the conviction that Lady Gregory's 'Cuchulain of Muirthemne', though only a paraphrase, gave the best idea of the Ulster stories. This merely emphasised the dearth, for her book, even as a paraphrase, seemed lacking in some important ways, refining away the coarse elements and rationalising the monstrous and gigantesque; as well as this, the *Táin Bó Cuailnge*, the prose epic which is the centre-piece of the Ulster cycle — and the oldest vernacular epic in Western literature — seemed inadequately represented.

The *Táin*, or Cattle Raid, is the nearest approach to a great epic that Ireland has produced. For parts of the narrative, and for some of the ancillary stories, achievements at the highest level of saga literature may fairly be claimed. It seemed extraordinary that, for all the romanticised, fairy tale, versified, dramatised and bowdlerised versions of the Ulster cycle, there had never been a readable translation of the older version of the *Táin*, tidied a little and completed from other sources — nothing in English to give an idea of the story as we first have it. So I undertook the present translation, and completed it as time offered. It is not intended as a scholarly work (for which I had neither motive nor equipment) but as a living version of the story, leaving as few obstacles as possible between the original and the reader.

Grateful acknowledgement is made to all who in any way helped with this translation, in particular

to Bórd Scoláireachtaí Cómalairte for a six-month fellowship in 1963, which enabled the main translation to begin; to the Minister for Finance for permission to take advantage of the fellowship; and to Professors James Carney and David Greene and the late Donagh Mac Donagh for their help in this matter;

to Southern Illinois University for the time and facilities to finish the translation;

to Professor John V. Kelleher of Harvard University, Professor Proinsias MacCana of University College Dublin and Mr. Gene C. Haley of Harvard University for the help detailed in the Introduction;

to Professor David Greene for his help in connection with some amendments to the first edition;

to Liam Miller for his endless forbearance and enthusiasm; and to my wife Eleanor for her help at all stages of the work.

T.K.

ARTIST'S NOTE

Any graphic accompaniment to a story which owes its existence to the memory and concern of a people over some twelve hundred years, should decently be as impersonal as possible.

The illuminations of early Celtic manuscripts express not personality but temperament. They provide not graphic comment on the text but an extension of it. Their means are not available to us today — either temperamentally or technically — but certain lessons may be learned from them relevant to the present work. In particular they suggest that graphic images, if any, should grow spontaneously and even physically from the matter of the printed text.

If these images — these marks in printer's ink — form an extension to Kinsella's *Táin*, they are a humble one. It is as shadows thrown by the text that they derive their substance.

L. le B.
Carros, France
November 1968

INTRODUCTION

MUCH of early Irish literature has been lost. Much of what survives is contained in a few large manuscripts made in medieval times. Among their miscellaneous contents are four groups of stories:

— mythological stories relating to the Tuatha Dé Danann ('the Tribes of the Goddess Danann') an ancient divine race said to have inhabited Ireland before the coming of the Celts;
— the Ulster cycle, dealing with the exploits of King Conchobor and the champions of the Red Branch, chief of whom is Cúchulainn, the Hound of Ulster;
— the Fenian cycle, stories of Finn mac Cumaill, his son Oisín, and the other warriors of the *fiana*; and
— a group of stories centred on various kings said to have reigned between the third century B.C. and the eighth century A.D.

The oldest of these manuscripts — *Lebor na hUidre*, familiarly known as the Book of the Dun Cow — was compiled in the monastery of Clonmacnoise in the twelfth century. It contains, in a badly flawed and mutilated text, part of the earliest known form of the *Táin Bó Cuailnge*. Another partial version of the same form of the story, also flawed, is contained in a late fourteenth century manuscript, the Yellow Book of Lecan. Between them these give the main body of the *Táin* as used in chapters II to XIV of this translation.

The origins of the *Táin* are far more ancient than these manuscripts. The language of the earliest form of the story is dated to the eighth century, but some of the verse passages may be two centuries older, and it is held by most Celtic scholars that the Ulster cycle, with the rest of early Irish literature, must have had a long oral existence before it received a literary shape, and a few traces of Christian colour, at the hands of the monastic scribes.

As to the background of the *Táin*, the Ulster cycle was traditionally believed to refer to the time of Christ. This might seem to be supported by the similarity between the barbaric world of the stories, uninfluenced by Greece or Rome, and the La Tène Iron Age civilisation of Gaul and Britain. The *Táin* and certain descriptions of Gaulish society by Classical authors have many details in common: in warfare alone, the individual weapons, the boastfulness and courage of the warriors, the practices of cattle-raiding, chariot-fighting and beheading. Ireland, however, by its isolated position, could retain traits and customs that had disappeared elsewhere centuries before, and it is possible that the kind of culture the *Táin* describes may have lasted in Ireland up to the introduction of Christianity in the fifth century.

The *Táin* itself, considered as a unit, lacks a number of essential elements: the actual motive for the Connacht invasion of Ulster, the reason for the sickness of the Ulster warriors throughout most of the action, the reason for Fergus's opposition to Conchobor and for the presence of a troop of Ulster exiles in the Connacht army.

The last three of these elements are supplied in separate tales. Many *remscéla*, or pre-tales, lead up to the *Táin*. Though not strictly part of the story they are important tributaries. Some tell of the origins, wooings and adventures of the kings and heroes of the cycle. The first section of this translation consists of a group of *remscéla* chosen for their contribution to an understanding of the plot and motivations of the *Táin*. Their sources are identified in the section of detailed notes beginning on page 255. A separate series, not represented in the translation, deals with Connacht's build-up of alliances for the war on Ulster and the arrangements for provisioning the armies.

The motive for the invasion, given in chapter I of the translation, is supplied from a later form of the *Táin*, four centuries younger, of which the most famous copy is contained in another major manuscript compiled in the twelfth century, the Book of Leinster.

The language of the Book of Leinster version of the *Táin* is dated to the twelfth century, and the story survives complete. The author or compiler was at pains to produce a consistent and integral narrative. Perhaps because of its completeness the Book of Leinster *Táin* has had considerable attention from editors and translators; among others, there is an abridged translation by Standish Hayes O'Grady in Eleanor Hull's 'The Cuchullin Saga,' published in 1898; the text, with a German translation, was published by Ernst Windisch in 1905; d'Arbois de Jubainville published a partial translation into French in 1907, and Joseph Dunn a full translation into English in 1914. It is also the version employed by Lady Gregory in 'Cuchulain of Muirthemne,' one of Yeats's source-books. Most recently, in 1967, the Dublin Institute for Advanced Studies published a new edition of the text, with an English translation, by Cecile O'Rahilly

Of the earlier version only one English translation has been published — by Winifred Faraday, in 1904 — but it is incomplete, and difficult to read with any pleasure, partly because it transmits the flaws of the text so accurately.

These might be thought sufficient grounds for choosing the *Lebor na hUidre*/Yellow Book of Lecan version for a new translation, but there is also strong reason in the actual quality of the earlier recension. The compiler of the Book of Leinster *Táin* had, besides a care for completeness, a generally florid and adjectival style, running at times to an overblown decadence. It has seemed better, despite the defects of the earlier text, to try to give an idea of the simple force of the story at its best.

This early text is the work of many hands and in places is little more than the mangled remains of miscellaneous scribal activities. There are major inconsistencies and repetitions among the incidents. On occasion the narrative withers away into cryptic notes and summaries. Extraneous matter is added, varying from simple glosses and comments to wholesale indiscriminate interpolations from other sources, in some cases over erased passages of the original; Frank O'Connor, in 'The Backward Look,' his short history of Irish literature, says that, as a result, 'the Cattle Raid has been rendered practically unintelligible'. Many of the defects can be remedied, however, and a reasonably coherent narrative extracted, with a little reorganisation. This the present translation has attempted to do, taking it as a principle that any interference with the text as it stands, even when it seems most necessary, should be kept to a minimum.

As far as possible the story has been freed of inconsistencies and repetitions. Obscurities have been cleared up and missing parts supplied from other sources, generally the Book of Leinster text, but this has been done as economically as possible, sometimes with only a word or phrase. The passages introduced by the compiler or interpolator, where they are not involved in this tidying process, have been left undisturbed; they will be readily recognised by the changes in style. Such matters are noted (it is hoped adequately) as they occur, together with any changes in sentence order or other similar amendments. A reader who is anxious to know how the text actually runs should be able to restore the original disarray.

Nothing has been added in the translation beyond a very occasional word or phrase designed to keep the narrative clear; these additions are noted. But there are two aspects of the translation not fully 'covered by guarantee.' The first has to do with the main purpose of the work, which is to give a readable and living version of the story: it is that no attempt has been made to preserve the actual texture of the Irish narrative. Sentence structure and tense, for example, have been changed without hesitation; elements are occasionally shifted from one sentence to another; proper names have been substituted for pronouns, and vice versa; a different range of verbs has been used; and so on. This is not, therefore, a literal translation. But it is a close compromise with one, and tries not to deviate significantly at any point from the original.

The second exception has to do with the verse passages: greater freedom has been taken with the verse than with the prose, though the sense and structural effects are followed with reasonable faithfulness. For one category of verse, however, the guarantee has to be withdrawn completely — the passages of *rosc* or *retoiric* which occur in 'stepped' form throughout the translation. In the original these are extremely obscure. This is partly because, as is generally believed, they are more archaic, but it seems likely that in some instances, where the utterance is 'deep' or prophetic, the obscurity is also deliberate. Scholars have preferred on the whole to leave these verses unattempted, but it seemed worthwhile to try to make some sense out of them, especially where something central to the action is going on, as in the long sequence in chapter VI. The aim has been to produce passages of verse which more or less match the original for length, ambiguity and obscurity, and which carry the phrases and motifs and occasional short runs that are decipherable in the Irish. It is stressed that they are highly speculative and may reproduce little if anything of the original effect. It would have been impossible to attempt these *rosc* passages without expert help, and for most of those in the *Táin* itself this was given with much tolerance, patience and generosity by Dr. Proinsias Mac Cana of University College, Dublin. His suggestions were offered as starting points for the imagination and they have undergone much violence in the process that followed, for which the translator is solely responsible. No attempt has been made to follow the Irish verse forms.

For a great deal of detailed information in a short space on the *Táin* and its background, the introduction to Cecile O'Rahilly's *Táin Bó Cúalnge* is recommended. For a proper consideration of the place of the *Táin* in saga literature, or in the Irish tradition, and for comments on the historical, mythological, symbolic, or other larger aspects, a list of further reading is suggested on page xxiv.

Scholars and commentators continue to pursue these topics with fascinating results. For Alwyn and Brinley Rees, for example, in *Celtic Heritage*, '. . . the *Táin* appears as an example of the classic struggle between the priestly and the warrior classes, each of which tends to usurp the functions of the other.' Heinrich Zimmer saw the pillow-talk which gives rise to the *Táin* as a conflict between Celtic-Aryan father-dominance and the mother-dominance of the pre-Celtic inhabitants of the British Isles. Frank O'Connor suggested that the earliest layer of the story, incompletely preserved in the *rosc* passages, constitutes the remains of an ancient ironic anti-feminist poem. T. F. O'Rahilly believed that the Ulster stories describe the historical circumstances of the invasion of Ulster by Uí Néill invaders from Leinster (not Connacht), the idea of Medb as queen of Connacht —'Medb' was in fact the tutelary goddess of Temair, or Tara, in Leinster — being a mistake on the part of writers who were unaware that Irish tribes did not have queens.

Discussion of these and similar matters is outside the scope of this book, but a few features of the *Táin* as it is presented here seem to call for comment. It is quickly apparent, for example, that even allowing for the later interpolations there is no unifying narrative tone: the story is told in places with a neutral realism, in places with an air of folk or fantasy. It is clear also that the verbal or conventional quality of the narrative counts occasionally for more than consistency: Finnabair, for example, who dies of shame in chapter XII, is brought back to life and 'stays with' Cúchulainn for the sake of a storyteller's flourish at the end of the *Táin*, a flourish that is factually inconsistent with that at the end of 'Cúchulainn's Courtship of Emer'. Lists of heroes, similarly, are very likely to contain some who (taking the story realistically) are dead or absent; the seven Maine, however many are killed, remain seven; hero after hero is reluctant to meet Cúchulainn because he is a beardless boy, despite the mounting evidence of his prowess; Cúchulainn vows to kill Medb whenever he sees her, but foregoes his chance at the end of the *Táin*, 'not being a killer of women'—though he has just previously killed the two lamenting women sent by Medb to deceive him (not to mention Lochu, in mistake for Medb, in chapter V and Eis Enchenn in 'Cúchulainn's Courtship of Emer'.) But these things are familiar in epic literature; to see them as 'mistakes', or even flaws, is to misapply modern conventions and ignore the real nature, and perhaps the oral origin, of the stories.

One of the major elements of the *Táin* is its topography. Place-names and their frequently fanciful meanings and origins occupy a remarkable place by modern standards. It is often enough justification for the inclusion of an incident that it ends in the naming of some physical feature; certain incidents, indeed, seem to have been invented merely to account for a place-name. The outstanding example is in the climax of the *Táin* itself, where the final battle (toward which we might assume the action is leading) is treated very casually, while attention is directed in detail to the wanderings of the mortally-wounded Donn Cuailnge around Ireland, naming the places as he goes.

This phenomenon is not confined to the *Táin*, or the Ulster cycle; it is a continuing preoccupation of early and medieval Irish literature, which contains a whole class of topographical works, including prose tracts and poems of enormous length composed by the professional poets, who were expected to recite from them on demand.

The topographical element is important for a full appreciation of the *Táin*. Much of the action consists of the movement of the Connacht armies across Ireland and back and forth over the country which makes up part of the present-day County Louth. We can be certain about the identification of some of the *Táin* place-names and certain also that others are unidentifiable, having been replaced by English names over the years, or being merely descriptive words or phrases that might apply to any number of places, or the result of a storyteller's or scribe's rationalisation or fantasy. But there are many names which, short of complete certainty, can be identified with reasonable confidence. The *Táin's* sense of place is on the whole much more realistic than its sense of place-names and their origin, and it is possible to follow the route of the *Táin* in all its essentials.

Maps of the route are given on pages xix - xxiii. Simplified as these maps are, it would have been impossible to put them together without expert help. This was provided by Professor John V. Kelleher and Mr. Gene C. Haley of Harvard University, who have been extraordinarily generous with the fruit of their researches, and with many suggestions. Their findings however, are not yet final, and are by no means adequately represented here. The maps in the present work are designed only to give enough information to follow the movements of the story.

A strong element in the sagas is their directness in bodily matters: the easy references to seduction, copulation, urination, the picking of vermin, the suggestion of incest in 'How Cúchulainn was Begotten', and so on. This coarseness was a source of some uneasiness to Lady Gregory ('I left out a good deal I thought you would not care about for one reason or another', she wrote to the people of Kiltartan), but it seems very mild to a modern reader — an effect of the same directness with which the story treats killing and mutilation. If Fergus's adultery with Medb calls for special comment it is only because their relationship is of central importance in the *Táin*, from the clash over the Galeóin at the beginning of the expedition to the true curtain-line uttered by Fergus at the end of the great battle. Their encounter in the wood, where Fergus (clearly not up to Medb's demands) loses his sword, is the source of continual phallic joking in the *Táin* until the sword is restored before the battle and Fergus goes out to pour his released fury on Conchobor, the old usurper.

Probably the greatest achievement of the *Táin* and the Ulster cycle is the series of women, some in full scale and some in miniature, on whose strong and diverse personalities the action continually turns: Medb, Derdriu, Macha, Nes, Aife. It may be as goddess-figures, ultimately, that these women have their power; it is certainly they, under all the violence, who remain most real in the memory.

The story, as the *Táin* ends, is not finished; other tales continue the action. The most important of these for the plot are:

Cath Ruis na Ríg, 'The Battle of Ros na Ríg', telling of Ulster's war of revenge for the *Táin*, and how Cúchulainn killed Coirpre, king of Temair;

Aided Con Roi, 'The Death of Cúroi', telling how Cúchulainn treacherously murdered Cúroi after being shamed in battle by him;

Brislech mór Maige Muirtheimne, 'The great Carnage on Murtheimne Plain' (not to be confused with the episode of the same title in the *Táin Bó Cuailnge*) and *Aided ConCulainn*, 'The Death of Cúchulainn', telling of Ulster's defeat at the hands of her united enemies, and how Cúchulainn was killed by the sons of Coirpre, Cúroi and Calatín (the Gaile Dána of the earlier version of the *Táin*); and

Dergruather Chonaill Chernaigh, 'Conall Cernach's Red Onslaught', telling how Conall Cernach avenged Cúchulainn's death.

But these tales, though they bring the action further, do so in a very different mode, one that is characterised by high fantasy and a free recourse to the supernatural, so that a sequel to this translation would be a very different kind of book. These stories are also much inferior in quality on the whole to the earlier *Táin* and the best of the *remscéla;* in them the faults of slackness and bombast evident in the later *Táin* are magnified to a high degree.

There is another group of tales — among them some of the best in the Ulster cycle — which deal with exploits not directly related to the *Táin*. Of these the most important are:

Serglige ConCulainn ocus aenét Emire, 'Cúchulainn's Sickbed and Emer's One Jealousy', telling of the goddess Fann's love for Cúchulainn and how she lured him to the underworld;

Fled Bricrenn, 'Bricriu's Feast', and *Scél mucce Maic Dathó*, 'The Story of Mac Dathó's pig', telling of the struggles for precedence among the Ulster heroes;

Tochmarc Ferbe, 'The Courtship of Ferb', telling how Maine Mórgor, a son of Medb and Ailill, courted Gerg's daughter, and how Conchobor attacked the wedding-feast and killed Gerg and Maine and their followers;

a group of 'Death Tales' of the Ulster heroes;

Togail bruidne Da-Choca, 'The Ruin of Da-Choca's hostel', telling of the struggles for the succession after Conchobor's death; and

Siaburcharput ConCulainn, 'Cúchulainn's Demon Chariot', telling how Cúchulainn's spirit was called up by St. Patrick to help in the conversion of Laegaire, king of Ireland, to Christianity.

MAPS

Maps II and III in the following section indicate the route of the Connacht armies as described in the *Táin*. The itinerary on pages 63/65 of the translation matches this route only in parts, and has been ignored in preparing the maps.

The maps in this edition have been specially drawn by Don Farrell. No published map could be found which would give a general idea of the nature of the terrain and at the same time show enough of the present-day towns and other features to enable the reader to associate the route of the *Táin* with its modern setting; a list of equivalents is given, therefore, overleaf. The events in Cuailnge may be followed on the Irish Ordnance Survey one inch map No. 71.

In the maps, as in the translation, the names of the provinces are given in their modern form and all other names in their old Irish form, standardised by reference to Thurneysen's list in *Die irische Helden- und Königsage*

County Roscommon

Cruachan : Rathcroghan.

Cúil Silinne : in the townland of Ardakillan.

County Longford

Gránaird : Granard.

County Meath

Iraird Cuillenn : Crossakeel.

Cúil Sibrille : Kells.

Ath Gabla : a ford on the Mattock, near Kellystown.

County Louth

the river Níth : the Castletown river.

Ath Lethan : Dundalk Harbour.

Ath Carpat : the Big Bridge, in Dundalk.

Cuinciu : Slievenaglogh.

Finnabair Chuailnge : the High Rath, north of The Bush in the Carlingford peninsula.

the river Cronn : the Big River, in the Carlingford peninsula.

Bernas Bó Cuailnge : the Windy Gap, above the source of the Big River.

Glenn Dáilimda : the valley above Omeath.

Bélat Ailiuin : a passage over the Flurry River at Ravensdale

Liasa Liac : probably at Ballymakellett.

Glenn Gatlaig : the valley above Ballymakellett.

Dubchoire : a 'black cauldron', or recess, in the north side of Glenn Gatlaig.

Ochaine : Trumpet Hill.

Delga : an ancient fort west of Dundalk.

Focherd : Faughart.

Ath Da Ferta : a ford on the river Fane, at Knockbridge.

Muid Loga : the town of Louth.

Ath Firdia : Ardee.

Smirommair : Smarmore.

County Meath

Tailtiu : Teltown.

County Westmeath

Gáirech : Garhy.

Slemain Midi : Slanemore.

Ath Luain : Athlone.

ALBA

Dún Sobairce

ULSTER

Es Ruaid

Emain
Macha ◉

SLIAB
FUAIT

CUIB

Delga

CUAILNGE

IRRUS
DOMNANN

CRICH
ROIS

MAG
MURTHEIMNE

CONNACHT

MAG
Cruachan ◉
AI

Garad

TETHBA

Enloch

MIDE

Ath Luain

Brug na Bóinne

Temair

Luglochta
Logo

BREGA

Benn Etair

LEINSTER

SINANN

Síd ar Femen

Temair
Luachra

MUNSTER

SIUIR

Miles 0 10 20 30 40 50 60 70 80 90 100

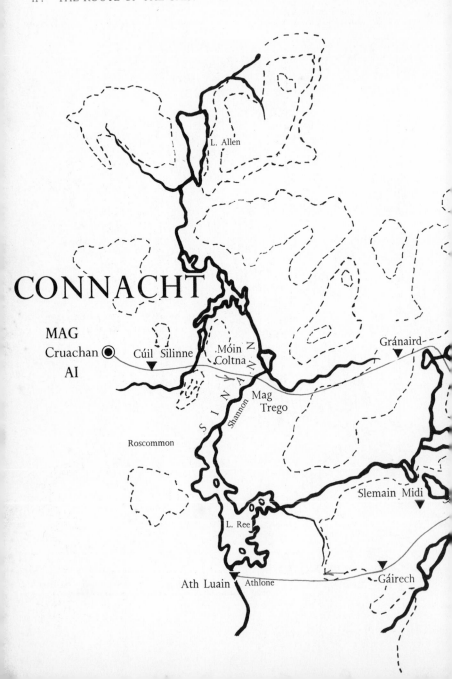

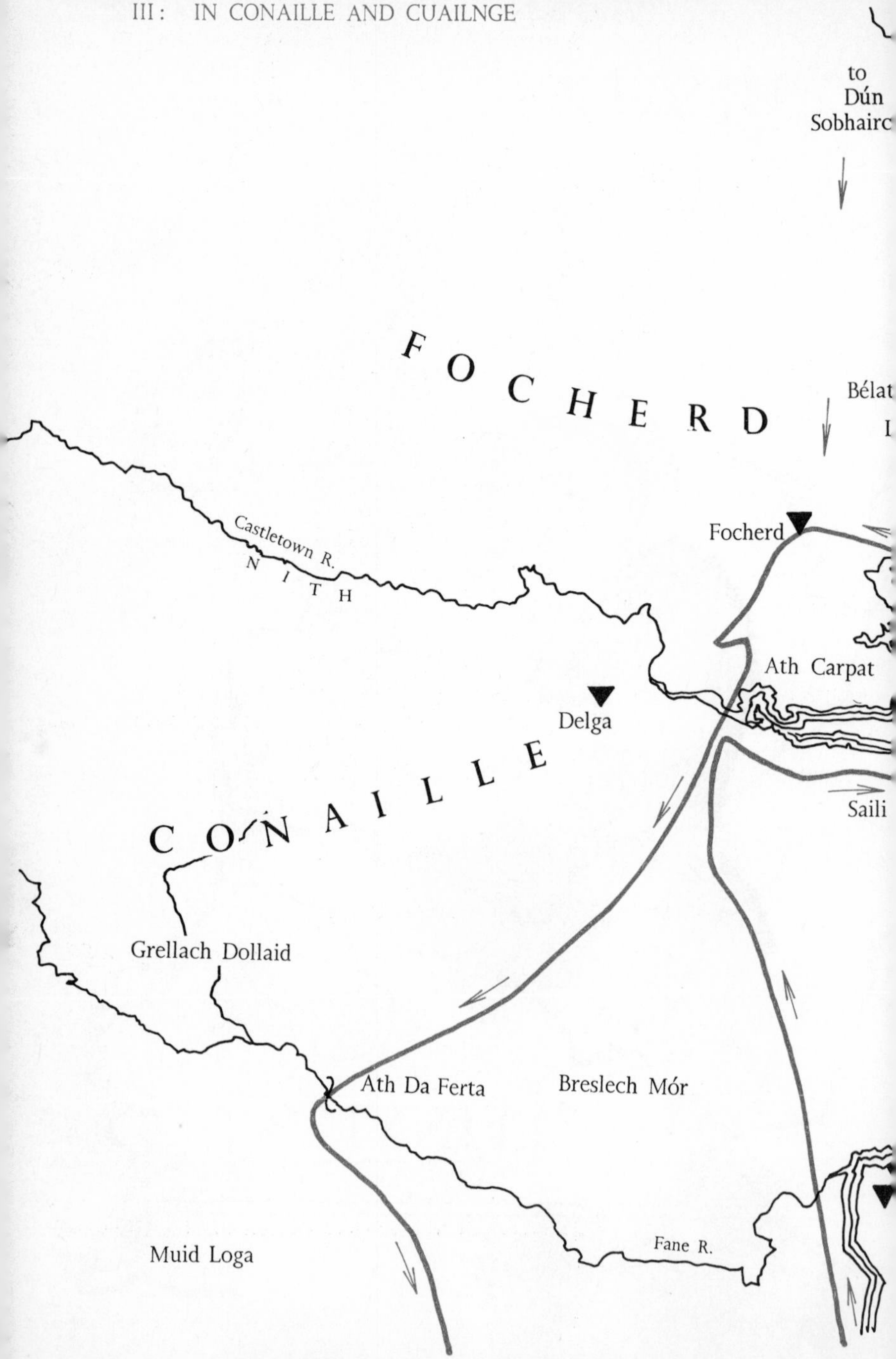

to
Dún
Sobhairc

F O C H E R D

Bélat
I

Castletown R.
N I T H

Focherd

Ath Carpat

Delga

Saili

C O N A I L L E

Grellach Dollaid

Ath Da Ferta   Breslech Mór

Muid Loga

Fane R.

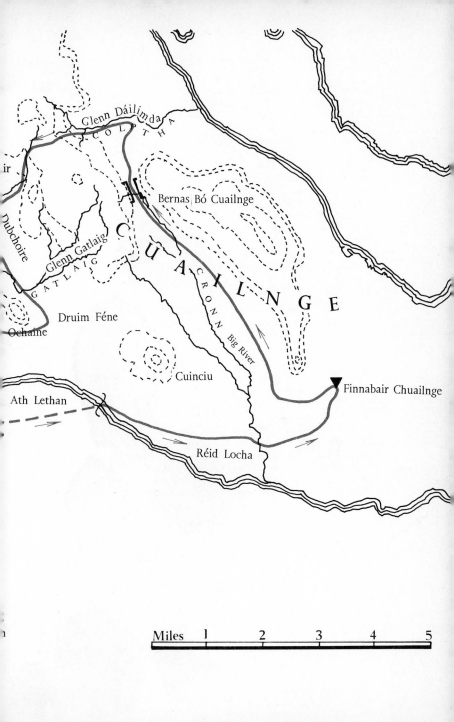

Glenn Dáilimda
COLOTHA
Bernas Bó Cuailnge
Dubchoire
Glenn Gatlaig
GATLAIG
C U A I L N G E
Druim Féne
Ochaine
CRONN Big River
Cuinciu
Finnabair Chuailnge
Ath Lethan
Réid Locha

Miles 1   2   3   4   5

SOME RECOMMENDED BOOKS

James Carney: *Studies in Irish Literature and History*, Dublin 1955.
Myles Dillon: *Early Irish Literature*, Chicago 1948.
Myles Dillon (ed.): *Irish Sagas* — Radio Eireann Thomas Davis Lectures, Dublin 1959.
Walter Fitzgerald: *The Historical Geography of Early Ireland*, London 1925.
Robin Flower: *The Irish Tradition*, Oxford 1947.
Eleanor Hull: *A Text Book of Irish Literature — Part I*, Dublin 1923.
Kuno Meyer and Alfred Nutt: *The Voyage of Bran*, London 1898.
Frank O'Connor: *The Backward Look — A Survey of Irish Literature*, London 1967.
T. F. O'Rahilly: *Early Irish History and Mythology*, Dublin 1946.
Alwyn Rees and Brinley Rees: *Celtic Heritage*, London 1961.
Marie-Louise Sjoestedt: *Gods and Heroes of the Celts*, London 1949.
Rudolf Thurneysen: *Die irische Helden- und Königsage* .., Halle 1921.

OTHER TRANSLATIONS

The Táin

Joseph Dunn: *The Ancient Irish Epic Tale Táin Bó Cúalnge*, London 1914.
Winifred Faraday: *The Cattle-Raid of Cualnge*, London 1904.
Cecile O'Rahilly: *Táin Bó Cúalnge from the Book of Leinster*, Dublin 1967.
Ernst Windisch: *Die altirischen Heldensage Táin Bó Cúalnge nach dem Buch von Leinster* . . . , Leipzig 1905.

General

Tom Peete Cross and Clark H. Slover: *Ancient Irish Tales*, London 1937.
Augusta Gregory: *Cuchulain of Muirthemne*, London 1902.
Eleanor Hull: *The Cuchullinn Saga in Irish Literature*, London 1898.
Patrick W. Joyce: *Old Celtic Romances*, London 1879.
A. H. Leahy: *Heroic Romances of Ireland*, London 1905 - 6.
Rudolf Thurneysen: *Sagen aus dem alten Irland*, Berlin 1901.

PRONUNCIATION OF IRISH WORDS

Reasonable English phonetic equivalents for the various proper names might have been used in the translation; the original orthography has been preferred, however, even though many of the words sound very odd pronounced as in English.

The following guide to pronunciation is very rough, and subject to exceptions, but may be some help.

consonants

Consonants at the beginning of a word have the same value as in English (c always = k). Elsewhere

b = v
c = g or k
ch is guttural, as in German
d = *dh* (as in 'then')
g = *gh*, a soft guttural
m = v (Fedelm and Leborcham are exceptions)
t = d
s followed or preceded by e or i = sh
th = th, as in 'thin'.

vowels

Short vowels have the same value as in Latin. Long vowels (marked with an accent) are pronounced awe, ay, ee, owe, oo (accents are not shown on capitals).

ai = a in the first syllable (Ailill, Maine) and i elsewhere (Cúchulainn, Emain)
a final e is sounded (Laegaire, Maine)
iu = u but with the i slightly sounded (Derdriu)
ui = i but with the u slightly sounded (Cuillenn)
ei = e (Deichtine, Murtheimne)

A pronunciation is suggested overleaf for some of the more important proper names in the stories; the stressed syllables are printed in bold type:

persons

Aife :	**ee** - fe
Ailill :	**al** - il
Amargin :	**av** - ar - *gh*in
Badb :	ba*dh*v/bive
Bricriu :	**brik** - ru
Cathbad :	ka*th* - va*dh*/**kaff** - a
Conchobor :	**kon** - *ch*ov - or/**kon** - *ch*or
Cúchulainn :	koo - *ch*ull - in
Deichtine :	de*ch* - tin - e
Derdriu :	**der** - dru
Dubthach :	**duv** - tha*ch*/**duff** - a*ch*
Emer :	**ay** - ver
Etarcomol :	**ed** - ar - **kov** - ol
Eochaid :	(y)**och** - i
Fedelm :	fe*dh* - elm
Fedlimid :	fe*dh* - lim - i*dh*/**fe** - lim - i
Ferdia :	fer - **di** - a
Finnabair :	**fin** - av - ir
Galeóin :	**gal** - (y)owe - in
Laeg :	loy*gh*
Leborcham :	**le** - vor - *ch*am
Laegaire :	loy*gh* - i - re
Lugaid :	loo*gh* - i*dh*/**loo** - i
Mac Roth :	mok - ro*th*
Maine :	**ma** - ne
Maine Athramail :	**ma** - ne a*th* - ra - vil
Medb :	me*dh*v/mayv
Morrígan :	**mo** - ree - *gh*an
Nemain :	**nev** - in
Noisiu :	**noy** - shu
Scáthach :	**skaw** - tha*ch*
Sétanta :	**shay** - dan - da
Sualdam :	**soo** - al - dav
Uisliu :	**ish** - lu

places, and other

Ath :	awth
Ath Ferdia :	**awth** ir - **di** - a
Ath Gabla :	**awth gav** - la
Ath Lethan :	**awth leth** - an
Benn Etair :	ben **ayd** - ir
Brug :	broo*gh*/broo
Colptha :	**kolp** - tha
Conaille :	**kon** - il - e
Craeb Ruad :	**krayv roo** - a
Cruachan Aí :	**kroo** - a - *cha*n **ee**
Cuailnge :	**koo** - ling - e
Cuib :	k(w)iv
Dub :	duv
Dún Sobairche :	**doon sov** - ir - *che*
Emain Macha :	**ev** - in **ma** - *cha*
Femen :	**fev** - en
Fid :	fi*dh*/fee
Iraird Cuillenn :	**ir** - ard **k(w)il** - en
Lugmod :	**loo** - vo*dh*
Mag :	ma*gh*
Mag Aí :	ma*gh* **ee**
Mag mBreg :	ma*gh* **mray**
Mag Murtheimne :	ma*gh* **mur** - thev - ne
Midi :	mi*dh* - i
Midluachair :	mi*dh* - **loo** - a - *chi*r
Síd :	shee*dh*
Sinann :	**shin** - an
Siuir :	**shoo** - ir
Sliab :	**shlee** - av
Sliab Fuait :	**shlee** - av **foo** - id
Sliab Mondairn :	**shlee** - av **mon** - dirn
Táin :	toyn
Tethba :	**teth** - va/**teff** - a

BEFORE THE TAIN

HOW THE TAIN BO CUAILNGE WAS FOUND AGAIN

THE POETS of Ireland one day were gathered around Senchán Torpéist, to see if they could recall the 'Táin Bó Cuailnge' in its entirety. But they all said they knew only parts of it.

Senchán asked which of his pupils, in return for his blessing, would travel to the land of Letha to learn the version of the Táin that a certain sage took eastward with him in exchange for the book Cuilmenn. Emine, Ninéne's grandson, set out for the east with Senchán's son Muirgen. It happened that the grave of Fergus mac Roich was on their way. They came upon the gravestone at Enloch in Connacht. Muirgen sat down at Fergus's gravestone, and the others left him for a while and went looking for a house for the night.

Muirgen chanted a poem to the gravestone as though it were Fergus himself. He said to it:

> 'If this your royal rock
>> were your own self mac Roich
> halted here with sages
>> searching for a roof
> *Cuailnge* we'd recover
>> plain and perfect Fergus.'

A great mist suddenly formed around him — for the space of three days and nights he could not be found. And the figure of Fergus approached him in fierce majesty, with a head of brown hair, in a green cloak and a red-embroidered hooded tunic, with gold-hilted sword and bronze blunt sandals. Fergus recited him the whole Táin, how everything had happened, from start to finish. Then they went back to Senchán with their story, and he rejoiced over it.

However, there are some who say that the story was told to Senchán himself after he had gone on a fast to certain saints of the seed of Fergus. This seems reasonable.

There are seven tales that prepare for the Táin Bó Cuailnge:

How Conchobor was Begotten, and how he took the Kingship of Ulster

The Pangs of Ulster

Exile of the Sons of Uisliu

How Cúchulainn was Begotten

Cúchulainn's Courtship of Emer and his Training in Arms with Scáthach

The Death of Aife's One Son

The Quarrel of the Two Pig-keepers and how the Bulls were Begotten

Some say that the tales before the Táin should also include Cúchulainn's coming to the house of Culann the smith, Cúchulainn's taking up arms and mounting into his chariot, and Cúchulainn's journey to the boy-troop in Emain Macha. But these three tales are given in the body of the Táin.

HOW CONCHOBOR WAS BEGOTTEN, AND HOW HE TOOK THE KINGSHIP OF ULSTER

NES the daughter of Eochaid Sálbuide of the yellow heel was sitting outside Emain with her royal women about her. The druid Cathbad from the Tratraige of Mag Inis passed by, and the girl said to him:

'What is the present hour lucky for?'

'For begetting a king on a queen,' he said.

The queen asked him if that were really true, and the druid swore by god that it was: a son conceived at that hour would be heard of in Ireland for ever. The girl saw no other male near, and she took him inside with her.

She grew heavy with a child. It was in her womb for three years and three months. And at the feast of Othar she was delivered.

The boy Conchobor was reared by Cathbad and was known as Cathbad's son. Conchobor rose to great dignity seven years after his birth, when he took the kingship of Ulster This is how it happened.

His mother Nes was living by herself. Fergus mac Roich was king in Ulster at that time, and he sought Nes for his wife.

'Only if I get something in return,' she said. 'Give my son the kingship for a year, so that they can call his son the son of a king.'

'Let her have that,' everyone said. 'You'll still be king, even though we call him by the name of king.'

So the woman slept with Fergus, and Conchobor was called 'king of Ulster.' Nes immediately set about advising her son and his foster-parents and everyone in his household. They were to steal everything from one half of the people and give it away to the other half. She gave the Ulster warriors her own gold and silver — all this was in hopes of what her son would get.

In a year to the same day the time was up. Fergus called back his pledge.

'We'll have to talk about this,' the men of Ulster said.

They gathered together and talked : they felt greatly insulted that Fergus had given them over, like a dowry, while they were grateful to Conchobor for all he had given them. They decided, 'What Fergus sold let it stay sold; what Conchobor bought let it stay bought.'

In this way Fergus parted with the kingship of Ulster and Cathbad's son Conchobor became high king of a province of Ireland.

Ulster grew to worship Conchobor. So high was their regard for him that every man in Ulster that took a girl in marriage let her sleep the first night with Conchobor, so as to have him first in the family. There was no wiser being in the world. He never gave a judgment until it was ripe, for fear it might be wrong and the crops worsen. There was no harder warrior in the world, but because he was to produce a son they never let him near danger. Heroes and battle-veterans and brave champions went before him into every fight and fray, to keep him from harm. Any Ulsterman who gave him a bed for the night gave him his wife as well to sleep with.

His household was very handsome. He had three houses: Craebruad, the Red Branch; Téte Brec, the Twinkling Hoard; and Craebderg, the Ruddy Branch. The severed heads and spoils were kept in the Craebderg. The kings sat in the Craebruad, red being for royalty. All the javelins and shields and swords were kept in the Téte Brec; the place twinkled with the gold of sword-hilts and the gold and silver glimmering on the necks and coils of grey javelins, on shield-plates and shield-rims, and in the sets of goblets, cups and drinking-horns.

Ochain was there, Conchobor's shield, the Ear of
Beauty — it had four gold borders around it;
Cúchulainn's black shield Dubán;
Lámthapad — the swift to hand — belonging to Conall
Cernach;
Ochnech belonging to Flidais;
Furbaide's red-gold Orderg;
Cúscraid's triumphant sword Coscrach;
death-dealing Echtach that belonged to Amargin;
Condere's angry Ir;
Nuadu's Cainnel — a bright torch;
Fergus's hacking sword Leochain;
the fearful Uathach that belonged to Dubthach;
Errge's Lettach;
Menn's Brattach;
Noisiu's joyful Luithech;
Nithach the wounder belonging to Laegaire;
the bloody Croda of Cormac;
Sencha's resonant shield Sciatharglan;
Celtchar's Comla Catha, the Door of Battle;
and other shields beyond counting.

Also beyond counting were Conchobor's household and his houses. There were one hundred and fifty inner rooms, each of which held three couples. The houses and rooms were panelled with red yew. In the centre of the house was Conchobor's own room, guarded by screens of copper, with bars of silver and gold birds on the screens, and precious jewels in the birds' heads for eyes. Over Conchobor's head was a rod of silver with three apples of gold, for keeping order over the throng. If it shook, or he raised his voice, everyone fell into such a respectful silence you would hear a needle drop to the floor. At any given time in Conchobor's room there were thirty noble heroes drinking out of Gerg's vat, which was always kept full. This was Ol nguala, the 'coal vat' that Conchobor took with him from Gerg's Glen when he killed Gerg.

THE PANGS OF ULSTER

What caused the pangs of the men of Ulster?
It is soon told.

THERE was a very rich landlord in Ulster, Crunniuc mac Agnomain. He lived in a lonely place in the mountains with all his sons. His wife was dead. Once, as he was alone in the house, he saw a woman coming toward him there, and she was a fine woman in his eyes. She settled down and began working at once, as though she were well used to the house. When night came, she put everything in order without being asked. Then she slept with Crunniuc.

She stayed with him for a long while afterward, and there was never a lack of food or clothes or anything else under her care.

Soon, a fair was held in Ulster. Everyone in Ulster, men and women, boys and girls, went to the fair. Crunniuc set out for the fair with the rest, in his best clothes and in great vigour.

'It would be as well not to grow boastful or careless in anything you say,' the woman said to him.

'That isn't likely,' he said.

The fair was held. At the end of the day the king's chariot was brought onto the field. His chariot and horses won. The crowd said that nothing could beat those horses.

'My wife is faster,' Crunniuc said.

He was taken immediately before the king and the woman was sent for. She said to the messenger:

'It would be a heavy burden for me to go and free him now. I am full with child.'

'Burden?' the messenger said. 'He will die unless you come.

She went to the fair, and her pangs gripped her. She called out to the crowd:

'A mother bore each one of you! Help me! Wait till my child is born.'

But she couldn't move them.

'Very well,' she said. 'A long-lasting evil will come out of this on the whole of Ulster.'

'What is your name?' the king said.

'My name, and the name of my offspring,' she said, 'will be given to this place. I am Macha, daughter of Sainrith mac Imbaith.'

Then she raced the chariot. As the chariot reached the end of the field, she gave birth alongside it. She bore twins, a son and a daughter. The name Emain Macha, the Twins of Macha, comes from this. As she gave birth she screamed out that all who heard that scream would suffer from the same pangs for five days and four nights in their times of greatest difficulty. This affliction, ever afterward, seized all the men of Ulster who were there that day, and nine generations after them. Five days and four nights, or five nights and four days, the pangs lasted. For nine generations any Ulsterman in those pangs had no more strength than a woman on the bed of labour. Only three classes of people were free from the pangs of Ulster: the young boys of Ulster, the women, and Cúchulainn. Ulster was thus afflicted from the time of Crunniuc, the son of Agnoman, son of Curir Ulad, son of Fiatach mac Urmi, until the time of Furc, the son of Dallán, son of Mainech mac Lugdach. (It is from Curir Ulad that the province and people of Ulster — Ulad –– have their name.)

EXILE OF THE SONS OF UISLIU

What caused the exile of the sons of Uisliu?
It is soon told.

THE MEN of Ulster were drinking in the house of Conchobor's storyteller, Fedlimid mac Daill. Fedlimid's wife was overseeing everything and looking after them all. She was full with child. Meat and drink were passed round, and a drunken uproar shook the place. When they were ready to sleep the woman went to her bed. As she crossed the floor of the house the child screamed in her womb and was heard all over the enclosure. At that scream every-

one in the house started up, ready to kill. Sencha mac
Ailella said:
   'No one move! Bring the woman here. We'll see what
caused this noise.'
   So the woman was brought before them. Her husband
Fedlimid said:
>   'Woman,
>   what was that fierce shuddering sound
>   furious in your troubled womb?
>   The weird uproar at your waist
>   hurts the ears of all who hear it.
>   My heart trembles at some great terror
>   or some cruel injury.'

She turned distracted to the seer Cathbad:
>   'Fair-faced Cathbad, hear me
>   — prince, pure, precious crown,
>   grown huge in druid spells.
>   I can't find the fair words
>   that would shed the light of knowledge
>   for my husband Fedlimid,
>   even though it was the hollow
>   of my own womb that howled.
>   No woman knows what her womb bears.'

Then Cathbad said:
>   'A woman with twisted yellow tresses,
>   green-irised eyes of great beauty
>   and cheeks flushed like the foxglove
>   howled in the hollow of your womb.
>   I say that whiter than the snow
>   is the white treasure of her teeth;
>   Parthian-red, her lip's lustre.
>   Ulster's chariot-warriors
>   will deal many a blow for her.

There howled in your troubled womb
a tall, lovely, long-haired woman.
Heroes will contend for her,
high kings beseech on her account;
then, west of Conchobor's kingdom
a heavy harvest of fighting men.
High queens will ache with envy
to see those lips of Parthian-red
opening on her pearly teeth,
and see her pure perfect body.'

Cathbad placed his hand on the woman's belly and
the baby wriggled under it.

'Yes,' he said, 'there is a girl there. Derdriu shall be
her name. She will bring evil.'

Then the daughter was born and Cathbad said:

'Much damage, Derdriu, will follow
your high fame and fair visage:
Ulster in your time tormented,
demure daughter of Fedlimid.

And later, too, jealousy
will dog you, woman like a flame,
and later still — listen well —
the three sons of Uisliu exiled.

Then again, in your lifetime,
a bitter blow struck in Emain.
Remorse later for that ruin
wrought by the great son of Roech;

Fergus exiled out of Ulster
through your fault, fatal woman,
and the much-wept deadly wound
of Fiachna, Conchobor's son.

Your fault also, fatal woman,
Gerrce felled, Illadan's son,
and a crime that no less cries out,
the son of Durthacht, Eogan, struck.

Harsh, hideous deeds done
in anger at Ulster's high king,
and little graves everywhere
— a famous tale, Derdriu.'

'Kill the child!' the warriors said.

'No,' Conchobor said. 'The girl will be taken away tomorrow. I'll have her reared for me. This woman I'll keep to myself.'

The men of Ulster didn't dare speak against him.

And so it was done. She was reared by Conchobor and grew into the loveliest woman in all Ireland. She was kept in a place set apart, so that no Ulsterman might see her until she was ready for Conchobor's bed. No one was allowed in the enclosure but her foster-father and her foster-mother, and Leborcham, tall and crooked, a satirist, who couldn't be kept out.

One day in winter, the girl's foster-father was skinning a milk-fed calf on the snow outside, to cook it for her. She saw a raven drinking the blood on the snow. She said to Leborcham:

'I could desire a man who had those three colours there: hair like the raven, cheeks like blood and his body like snow.'

'Good luck and success to you!' Leborcham said. 'He isn't too far away, but close at hand — Noisiu, Uisliu's son.'

'I'll be ill in that case,' she said, 'until I see him.'

This man Noisiu was chanting by himself one time near Emain, on the rampart of the stronghold. The chanting of the sons of Uisliu was very sweet. Every cow or beast that heard it gave two thirds more milk. Any person hearing it was filled with peace and music. Their deeds in war were great also: if the whole province of Ulster came at them at once, they could put their three backs together and not be beaten, their parrying and defence were so fine. Besides this they were swift as hounds in the chase, killing the wild beasts in flight.

While Noisiu was out there alone, therefore, she slipped out quickly to him and made as though to pass him and not recognize him.

'That is a fine heifer going by,' he said.

'As well it might,' she said. 'The heifers grow big where there are no bulls.'

'You have the bull of this province all to yourself,' he said, 'the king of Ulster.'

'Of the two,' she said, 'I'd pick a game young bull like you.'

'You couldn't,' he said. 'There is Cathbad's prophecy.'

'Are you rejecting me?'

'I am,' he said.

Then she rushed at him and caught the two ears of his head.

'Two ears of shame and mockery,' she said, 'if you don't take me with you.'

'Woman, leave me alone!' he said.

'You will do it,' she said, binding him.

A shrill cry escaped him at that. The men of Ulster nearby, when they heard it, started up ready to kill. Uisliu's other sons went out to quieten their brother.

'What is wrong?' they said. 'Whatever it is, Ulstermen shouldn't kill each other for it.'

He told them what had happened.

'Evil will come of this,' the warriors said. 'But even so, you won't be shamed as long as we live. We can bring her with us to some other place. There's no king in Ireland who would deny us a welcome.'

They decided on that. They left that night, with three times fifty warriors and three times fifty women and the same of hounds and menials. Derdriu was among them, mingling with the rest.

They travelled about Ireland for a long time, under protection. Conchobor tried to destroy them often with ambushes and treachery. They went round southwest-ward from the red cataract at Es Ruaid, and to the promontory at Benn Etair, northeastward. But still the men of Ulster pursued them until they crossed the sea to the land of Alba.

They settled there in the waste places. When the mountain game failed them they turned to take the people's cattle. A day came when the people of Alba went out to destroy them. Then they offered themselves to the king of Alba, who accepted them among his people as hired soldiers. They set their houses on the green. They built their houses so that no one could see in at the girl in case there might be killing on her account.

It happened that a steward came looking around their house early one morning. He saw the couple sleeping. Then he went and woke the king:

'I never found a woman fit for you until today,' he said. 'There is a woman with Noisiu mac Uislenn who is fit for a king over the Western World. If you have Noisiu killed, you can have the woman to sleep with,' the steward said.

'No,' the king said, 'but go and ask her every day in secret.'

He did this, but every day he came she told Noisiu about it that night. Since nothing could be done with her, the sons of Uisliu were ordered into all kinds of traps and dangerous battles to have them killed. But they were so hard in the carnage that nothing came of it.

They tried her one last time. Then the men of Alba were called together to kill them. She told Noisiu this.

'Go away from here,' she said. 'If you don't leave here this night, you will be dead tomorrow.'

So they left that night and reached an island in the sea.

This news reached Ulster.

'Conchobor,' everyone said, 'it would be shameful if the sons of Uisliu fell in enemy lands by the fault of a bad woman. Better to forgive and protect them — to save their lives and let them come home — than for enemies to lay them low.'

'Let them come,' Conchobor said. 'Send for them, with guarantees of safety.'

This news was brought to them.

'It is welcome,' they said. 'We'll go if Fergus comes as a pledge of safety, and Dubthach and Conchobor's son Cormac.'

Then they went down with the messengers to the sea.

So they were brought back to Ireland. But Fergus was stopped through Conchobor's cunning. He was invited to a number of ale feasts and, by an old oath, couldn't refuse. The sons of Uisliu had sworn they would eat no food in Ireland until they ate Conchobor's food first, so they were bound to go on. Fiacha, Fergus's son, went on with them, while Fergus and Dubthach stayed behind. The sons of Uisliu came to the green at Emain. Eogan mac Durthacht, king of Fernmag, was there: he had come to make peace with Conchobor, with whom he had long been at enmity. He had been chosen to kill them. Conchobor's hired soldiers gathered around him so that the sons of Uisliu couldn't reach him. They stood in the middle of the green. The women settled on the ramparts of Emain.

Eogan crossed the green with his men. Fergus's son came and stood at Noisiu's side. Eogan welcomed Noisiu with the hard thrust of a great spear that broke his back. Fergus's son grasped Noisiu in his two arms and pulled him down and threw himself across him, and Noisiu was finished off through Fergus's son's body. Then the slaughter broke out all over the green. No one left except by spike of spear or slash of sword. Derdriu was brought over to Conchobor and stood beside him with her hands bound at her back.

Fergus was told of this, and Dubthach and Cormac. They came at once and did mighty deeds. Dubthach killed Maine, Conchobor's son. Fiachna, son of Conchobor's daughter Fedelm, was killed with a single thrust. Fergus killed Traigthrén, Traiglethan's son, and his brother. Conchobor was outraged, and on a day soon afterward battle was joined between them, and three hundred among the men of Ulster fell. Before morning Dubthach had massacred the girls of Ulster and Fergus had burned Emain.

Then they went to Connacht, to Ailill and Medb — not that this was a home from home for Ulstermen, but that they knew these two would protect them. A full three thousand the exiles numbered. For sixteen years they made sure that weeping and trembling never died away in Ulster; there was weeping and trembling at their hands every single night.

She was kept a year by Conchobor. In that time she never
gave one smile, nor took enough food or sleep, nor lifted
up her head from her knees. If they sent musicians to her,
she would say this following poem:

'Sweet in your sight the fiery stride
of raiding men returned to Emain.
More nobly strode the three proud
sons of Uisliu toward their home:

Noisiu bearing the best mead
— I would wash him by the fire —
Ardán, with a stag or a boar,
Anle, shouldering his load.

The son of Nes, battle-proud,
drinks, you say, the choicest mead.
Choicer still — a brimming sea —
I have taken frequently.

Modest Noisiu would prepare
a cooking-pit in the forest floor.
Sweeter then than any meat
the son of Uisliu's, honey-sweet.

Though for you the times are sweet
with pipers and with trumpeters,
I swear today I can't forget
that I have known far sweeter airs.

Conchobor your king may take delight
in pipers and in trumpeters
— I have known a sweeter thing,
the three sons' triumphant song.

Noisiu's voice a wave roar,
a sweet sound to hear forever;
Ardán's bright baritone;
Anle, the hunter's, high tenor.

Noisiu : his grave-mound is made
and mournfully accompanied.
The highest hero — and I poured
the deadly drink when he died.

His cropped gold fleece I loved,
and fine form — a tall tree.
Alas, I needn't watch today,
nor wait for the son of Uisliu.

I loved the modest, mighty warrior,
loved his fitting, firm desire,
loved him at daybreak as he dressed
by the margin of the forest.

Those blue eyes that melted women,
and menaced enemies, I loved;
then, with our forest journey done,
his chanting through the dark woods.

I don't sleep now,
nor redden my fingernails.
What have I to do with welcomes?
The son of Indel will not come.

I can't sleep,
lying there half the night.
These crowds — I am driven out of my mind.
I can neither eat nor smile.

What use for welcome have I now
with all these nobles crowding Emain?
Comfortless, no peace nor joy,
nor mansion nor pleasant ornament.'

If Conchobor tried to soothe her, she would chant this
following poem:

'Conchobor, what are you thinking, you
that piled up sorrow over woe?
Truly, however long I live,
I can't spare you much love.

The thing most dear to me in the world,
the very thing I most loved,
your harsh crime took from me.
I won't see him till I die.

I feel his lack, wearily,
the son of Uisliu. All I see —
black boulders on fair flesh
so bright once among the others.

Red-cheeked, sweet as the river-brink;
red-lipped; brows beetle-black;
pearly teeth gleaming bright
with a noble snowy light.

His figure easiest to find,
bright among Alba's fighting-men
— a border made of red gold
matched his handsome crimson cloak.

A soft multitude of jewels
in the satin tunic — itself a jewel :
for decoration, all told,
fifty ounces of light gold.

He carried a gold-hilted sword
and two javelins sharply tipped,
a shield rimmed with yellow gold
with a knob of silver at the middle.

Fergus did an injury
bringing us over the great sea.
How his deeds of valour shrank
when he sold honour for a drink !

If all Ulster's warriors
were gathered on this plain, Conchobor,
I would gladly give them all
for Noisiu, son of Uisliu.

Break my heart no more today.
In a short while I'll be no more.
Grief is heavier than the sea,
if you were but wise, Conchobor.'

'What do you see that you hate most?' Conchobor said.
'You, of course,' she said, 'and Eogan mac Durthacht!'
'Go and live for a year with Eogan, then,' Conchobor
said.

Then he sent her over to Eogan.

They set out the next day for the fair of Macha. She
was behind Eogan in the chariot. She had sworn that two
men alive in the world together would never have her.

'This is good, Derdriu,' Conchobor said. 'Between me and Eogan you are a sheep eyeing two rams.'

A big block of stone was in front of her. She let her head be driven against the stone, and made a mass of fragments of it, and she was dead.

HOW CUCHULAINN WAS BEGOTTEN

CONCHOBOR and the nobles of Ulster were
at Emain. A flock of birds came to Emain Plain and ate
all the plants and grasses out of the ground, and the very
roots. The men of Ulster grew angry seeing their land
ruined, and got nine chariots ready the same day to chase
them away — they were practised hunters of birds.
Conchobor mounted the chariot with his sister, the
woman Deichtine; she drove the chariot for her brother.
The Ulster warriors, Conall and Laegaire and the others,
came in their chariots, and Bricriu with them.

The birds flew at will before them across Sliab Fuait,
and across Edmonn and Breg Plain — there were no dikes
or fences or stone walls in Ireland at that time, only the

open plain. Pleasant and lovely was the flight of the birds, and their song. There were nine scores of birds with a silver chain between each couple. Each score went in its own flight, nine flights altogether, and two birds out in front of each flight with a yoke of silver between them. Toward nightfall three birds separated out from the rest.

The men of Ulster pressed on until they reached Brug on the Boann river, and night overtook them there. It snowed heavily upon them, and Conchobor told his people to unyoke their chariots and start looking for a shelter. Conall and Bricriu searched about and found a solitary house, newly built. They went up to it and found a couple there and were made welcome. But when they returned to their people, Bricriu said it was useless to go there unless they brought their own food and set the table themselves — that even so it would be meagre enough. Nevertheless, they went there with all their chariots, and crowded with difficulty into the house. Soon they found the door of the store-room, and by their usual mealtime the men of Ulster were drunk with their welcome and in good humour.

Later, the man of the house told them his wife was in her birth-pangs in the store-room. Deichtine went in to her and helped her bear a son. At the same time a mare at the door of the house gave birth to two foals. The Ulstermen took charge of the baby boy and gave him the foals as a present, and Deichtine nursed him.

When morning came there was nothing to be seen eastward of the Brug — no house, no birds — only their own horses, the baby and the foals. They went back to Emain and reared the baby until he was a boy.

He caught an illness then, and died. And they made a lamentation for him, and Deichtine's grief was great at the loss of her foster-son. She came home from lamenting

him and grew thirsty and asked for a drink, and the drink
was brought in a cup. She set it to her lips to drink from
it and a tiny creature slipped into her mouth with the
liquid. As she took the cup from her lips she swallowed the
creature and it vanished.

She slept that night and dreamed that a man came
toward her and spoke to her, saying she would bear a
child by him — that it was he who had brought her to
the Brug to sleep with her there, that the boy she had
reared was his, that he was again planted in her womb
and was to be called Sétanta, that he himself was Lug
mac Ethnenn, and that the foals should be reared with the
boy.

The woman grew heavy with a child, and the people
of Ulster made much of not knowing its father, saying it
might have been Conchobor himself, in his drunkenness,
that night she had stayed with him at the Brug.

Then Conchobor gave his sister in marriage to Sualdam
mac Roich. She was ashamed to go pregnant to bed with
her husband, and got sick when she reached the bedstead.
The living thing spilled away in the sickness, and so she
was made virgin and whole and went to her husband.
She grew pregnant again and bore a son, and called him
Sétanta.

The men of Ulster were assembled in Emain Macha
when her son was born, and they began arguing over
which of them should rear the boy. They went to Conchobor for a decision.

'You should take the boy,' Conchobor said to his sister
Finnchaem.

Finnchaem looked at the boy.

'My heart is full of love for him already,' she said, 'as
though he were my own Conall Cernach.'

'It is only a little different for you,' Bricriu said; 'one your own son, and the other your sister's son.'

'Take the boy,' Conchobor said again to his sister.

'I should rear him,' Sencha said, 'and not Finnchaem. I am strong and skillful; I am noble and nimble in combat; I am a sage, knowing and careful. I have precedence over others in speaking with the king; I advise him before he speaks. I am judge of all combats before battle-proud Conchobor. I settle all judgments in Ulster, and offend no one. No one but Conchobor can equal me as a foster-father.'

'No: let me rear him,' Blai Briuga, a landed man, said. 'He will be safe from harm and neglect with me. I could take all the men of Ireland in my house, and feed them for a week, or ten days. In their rashness and wrath I sustain them. In times of insult and trials of honour I support them. But let my just claim be settled as Conchobor desires.'

'Have you no respect?' Fergus said. 'His wellbeing is my concern; I will rear him. I am strong and skillful, and a king's messenger. No one can match me for rank or riches. I am sharp in courage and the craft of arms. My honour is my constant care — I was made to mind foster-sons! I shelter the miserable, scourge the strong, watch over the weak.'

'Listen to me,' Amargin said, 'and don't turn away. I am worthy to bring up a king. I am renowned for every quality — for my deeds and wisdom and wealth, for eloquence and openness of mind, for the splendour and courage of my family. If I were not already a prince I would be worthy, as poet, of the royal favour. I can kill any chariot-fighter. I look for no one's thanks but Conchobor's. I am bound to no one but the king.'

'There is nothing to gain from this,' Conchobor said.

'Let Finnchaem have the boy until we reach Emain, and the judge Morann can decide.'

They set out for Emain, the boy with Finnchaem. When they reached Emain, Morann gave judgment and said:

'He should be given to Conchobor, for he is Finnchaem's kin. Sencha can teach him eloquence and oratory, Blai Briuga can provide for him, Fergus can take him on his knee, Amargin can be his teacher, with Conall Cernach as foster-brother. The teats of a mother Finnchaem can supply. In this manner he will be formed by all — chariot-fighter, prince and sage. He will be cherished by many, this boy, and he will settle your trials of honour and win your ford-fights and all your battles.'

That is what was done: he was given to Amargin and Finnchaem and reared at Imrith Fort on Murtheimne Plain.

CUCHULAINN'S COURTSHIP OF EMER, AND HIS
TRAINING IN ARMS

THE MEN of Ulster were with Conchobor in Emain Macha one time, drinking from the vat Ol nguala. It could hold a hundred measures of coal-black drink, enough to fill the men of Ulster for the whole evening at one sitting.

The Ulster chariot-warriors were practising on spear-cords stretched the length of the house from one door to the other, two hundred and five feet. The feats they performed were the apple-feat and the feats of the javelin and the sword-edge, and their names were Conall Cernach, the triumphant, son of Amargin; Fergus mac Roich, bravest of the brave; Laegaire Buadach, the victorious, son of Connad; Celtchar mac Uthidir; Dubthach mac Lugdach; Cúchulainn mac Sualdaim; and Scél, the son

of Bairdene who was doorkeeper of Emain Macha, and
after whom 'Bairdene's Pass' is named — Scél himself was
a great storyteller. Cúchulainn outdid them all by his
brilliance and nimbleness in the feats and the women of
Ulster filled with love for him, seeing him so brilliant,
clever and nimble of hand — and seeing also his fair face
and fine figure. Cúchulainn had no wife at that time, and
the men of Ulster met together to talk about him and
about their wives' and daughters' passion for him. They
said a woman would have to be found for Cúchulainn. A
man with a wife of his own would be less likely to ruin
their daughters and steal their wives' love. There was the
danger besides that Cúchulainn might die young and leave
no son, which would be tragic: they knew it was only
out of Cúchulainn himself that the like of him might
come again. For this reason also he should have a woman.

Conchobor sent nine men into each province of Ireland
looking for a woman for Cúchulainn. They looked in
every fort and every town of note in the country for a
king's daughter, or a noble's or landowner's, that Cúchulainn could take to wife. But after a year the messengers came back without a girl to suit him.

Cúchulainn himself went to a place called the Gardens
of Lug — Luglochta Logo — to woo a girl he knew there.
Her name was Emer and she was the daughter of Forgall
Monach, the cunning. Cúchulainn and his charioteer Laeg
mac Riangabra set out. No other chariot-team in Ulster,
horses or warriors, could touch that chariot, with those
warriors in it, for fire and speed.

Cúchulainn went up to the girl. She was out on the
green with her foster-sisters, the daughters of landowners
who lived around Forgall's fort. They were studying
embroidery and fine stitching with Emer.

Cúchulainn greeted the troop of girls and Emer lifted

up her lovely face. She recognised Cúchulainn, and said:
  'May your road be blessed!'
  'May the apple of your eye see only good,' he said.
  Then they spoke together in riddles.
  Cúchulainn caught sight of the girl's breasts over the top of her dress.
  'I see a sweet country,' he said. 'I could rest my weapon there.'
  Emer answered him by saying:
  'No man will travel this country until he has killed a hundred men at every ford from Scenmenn ford on the river Ailbine, to Banchuing — the 'Woman Yoke' that can hold a hundred — where the frothy Brea makes Fedelm leap.'
  'In that sweet country I'll rest my weapon,' Cúchulainn said.
  'No man will travel this country,' she said, 'until he has done the feat of the salmon-leap carrying twice his weight in gold, and struck down three groups of nine men with a single stroke, leaving the middle man of each nine unharmed.'
  'In that sweet country I'll rest my weapon,' Cúchulainn said.
  'No man will travel this country,' she said, 'who hasn't gone sleepless from Samain, when the summer goes to its rest, until Imbolc, when the ewes are milked at spring's beginning; from Imbolc to Beltine at the summer's beginning and from Beltine to Brón Trogain, earth's sorrowing autumn.'
  'It is said and done,' Cúchulainn said.
  He finished his journey and slept in Emain Macha that night.
  The girls told their fathers how the warrior came in his marvellous chariot, about the talk with its hidden mean-

ings that passed between him and Emer, and how he left
them, going northward across Breg Plain. The landowners
told it all to Forgall Monach, with everything that Emer
said.

'Plainly,' Forgall Monach said, 'it was the warped one
from Emain Macha. He came to talk with Emer, and the
girl fell in love with him. That is what the two of them
were talking about. But it will do them no good,' he said.
'I'll put an end to it. They'll never have what they want.'

Then Forgall Monach went to Emain Macha dressed
in Gaulish clothes. He said that royal messengers from
Gaul wished to speak with Conchobor, with tribute of
gold and Gaulish wine and other valuables. There were
three of them and they were given a great welcome. On
the third day he sent his people away.

Cúchulainn and Conall Cernach and other Ulster
chariot-warriors were highly praised in his presence. He
agreed they fought marvellously — though Cúchulainn,
if only he could visit Domnall Míldemail, the war-like,
in the land of Alba, would fight more marvellously still;
while if he visited Scáthach, the Shadowy One, and
studied the warrior's art with her, he could beat any hero
in Europe. This he suggested in the hope that Cúchulainn
would never come back, for he believed, if Cúchulainn
married Emer, that somehow through the warrior's
wildness and ferocity he himself would meet his death;
this was the source of his fear.

Cúchulainn said he would go and Forgall made him
promise to go immediately. Then, having got what he
wanted from Cúchulainn, he left and started for home.

Next morning the hero rose up and set out to fulfil his
promise. First he crossed Breg Plain to see Emer and talk
with her before he set sail. She told him it was Forgall
who had been in Emain and got him to go off studying

warfare, to keep them apart. She warned him to be on his guard, for Forgall would try to destroy him wherever he went. Each of them promised to stay pure until they met again, unless the other died. Then they took leave of each other and he turned toward Alba.

He stayed with Domnall and was taught first the Pierced Flagstone, with the bellows blowing under it. He performed on it until his soles were blackened and discoloured. Next the 'Hero's Coil on the Spikes of Spears' — climbing up along a spear and performing on its point without making his soles bleed.

Finally Domnall told Cúchulainn his training wouldn't be finished until he visited Scáthach further east in Alba. So he travelled across Alba.

Cúchulainn's road took him to the camp where Scáthach's pupils lodged. He asked where she might be.

'On that island there,' they said.

'How can I get to her?' he said.

'By the Pupils' Bridge,' they said. 'But no one can cross that unless he is trained in the craft of arms.' (It was made low at each end and high in the middle; no sooner did a person step on to one end but the other flew up at him and threw him on his back.)

Three times Cúchulainn tried to cross the bridge but his best efforts failed, and the men jeered him. Then he went into his warp-spasm. He stepped to the head of the bridge and gave his hero's salmon-leap onto the middle. He reached the far end of the bridge so quickly it had no time to fly up at him. Then he sprang off onto the solid ground of the island and went up to the fort. He struck the gate with his spear-point and broke through it.

Scáthach was told about this.

'Plainly,' she said, 'this is someone who has had his full training somewhere.'

She sent her daughter Uathach out to meet the young
man and see who he might be. Uathach saw him and fell
silent, his sweet shape woke such desire in her. She gazed
her fill at him and then went back to her mother. She
told her mother about the man she had seen, and praised
him.

'I can see he pleases you,' her mother said.

'Yes, indeed,' the girl said.

'Take him to bed tonight,' she said, 'and sleep with him,
if that is what you want.'

'It would be no hardship,' she said, 'if he would like to.'

The girl looked after him with water and food, pretend-
ing to be a servant and welcoming and entertaining him.
Later, Cúchulainn caught hold of her. But while he was
overpowering her he hurt her finger and she cried out.
Everyone in the fort heard her and they all started up.
Cochar Cruibne, one of Scáthach's soldiers and a very
hardened man, rushed at Cúchulainn and they struggled
and fought for a long time. Then Cochar tried his special
tricks of battle but Cúchulainn parried them as if he had
studied them all his life. Then he felled the champion
and cut his head off. The woman Scáthach mourned at
this, but Cúchulainn said he would take on the deeds and
duties of the dead man and lead her army and be her
strong champion.

Then Uathach came and conversed with Cúchulainn.
After three days the girl told Cúchulainn, if he really
wanted to learn heroic deeds, he must go where Scáthach
was teaching her two sons Cúar and Cat, and give his
hero's salmon-leap up to the big yew-tree where she was
resting, then put his sword between her breasts and make
her promise three things: thoroughness in his training, a
dowry for his marriage, and tidings of his future — for
Scáthach was also a prophetess.

Cúchulainn went up to Scáthach and stood with his feet on the two edges of the weapon-chest and stripped his sword and put the point to her heart and said:

'Death is hanging over you!'

'I'll give you any three things, the three highest desires of your heart,' she said, 'if you can ask them in one breath.'

Cúchulainn said what she had to give, and made her promise.

So Uathach lived with Cúchulainn and Scáthach taught him brave deeds and the craft of arms.

While Cúchulainn was staying with Scáthach in Alba and living with her daughter Uathach, another great man in Munster, a foster-brother of Cúchulainn, Lugaid mac Nois, who was the great king Alamiach's son, came eastward with twelve chariot-warriors, all princes of Munster, to woo the twelve daughters of Coirpre Niafer mac Rosa Ruaid. But the girls were all betrothed to men already. Forgall Monach heard of this and came to Temair. He told Lugaid he had an unmarried girl at home who was the finest in Ireland for shapeliness and purity and tidiness. Lugaid was glad to hear it, and Forgall promised the girl to him, and the Breg landowners' twelve daughters to Lugaid's twelve princes as well.

The king came to Forgall's fort for the wedding. Emer was brought to Lugaid's place to sit at his hand, but she held his cheeks and swore on his life and honour that it was Cúchulainn she loved, that she was under Cúchulainn's protection and that for anyone else to take her would be a crime against honour. Lugaid didn't dare sleep with Emer then for dread of Cúchulainn, and he turned for home.

At this period Scáthach was at war with another territory whose chief was the woman Aife. Their two armies gathered to give battle. Scáthach gave Cúchulainn a sleeping draught and tied him up, as a device to keep him out of battle, in case anything happened to him. But after one hour Cúchulainn sprang up straight from his sleep. A draught that would last another for twenty four hours lasted him only one.

He went out with Scáthach's two sons against three of Aife's soldiers, Cuar and Cat and Crufe, the three sons of Ilsúanach. Alone he reached the three of them and slew them. Battle was joined again next morning. The two armies came forward and lined up face to face. Three of Aife's soldiers, Ciri and Biri and Blaicne, sons of Eis Enchenn, the bird-headed, challenged Scáthach's two sons to combat. They chose the 'rope of feats' and Scáthach uttered a sigh, not knowing what would come of it. For one thing, her sons were only two men against three. For another, she dreaded Aife as the hardest woman warrior in the world. But Cúchulainn joined her two sons and sprang on to the cord and met the three and killed them.

Aife challenged Scáthach to single combat. Cúchulainn went up to Scáthach and asked her what Aife held most dear above all else.

'The things she holds most dear,' Scáthach said. 'are her two horses, her chariot and her charioteer.'

Cúchulainn met and fought Aife on the rope of feats. Aife smashed Cúchulainn's weapon. All she left him was a part of his sword no bigger than a fist.

'Look! Oh, look!' Cúchulainn said. 'Aife's charioteer and her two horses and the chariot have fallen into the valley! They are all dead!'

Aife looked round and Cúchulainn leaped at her and

seized her by the two breasts. He took her on his back like a sack, and brought her back to his own army. He threw her heavily to the ground and held a naked sword over her.

'A life for a life, Cúchulainn!' Aife said.

'Grant me three desires,' he said.

'What you can ask in one breath you may have,' she said.

'My three desires,' he said, 'are: hostages for Scáthach, and never attack her again; your company tonight at your own fort; and bear me a son.'

'I grant all you ask,' she said.

Cúchulainn went and slept the night with Aife.

Soon Aife said she was with child and would bear a boy.

'This day seven years I will send him to Ireland,' she said. 'But leave a name for him.'

Cúchulainn left him a gold thumb-ring and told her the boy was to come to Ireland to find him when his finger had grown to fit the ring. The name he gave him was Connla. He said Connla was to reveal this name to no man, that he must make way for no man, and refuse no man combat.

Then Cúchulainn went back to his own side.

He came back the way he had gone, and met a one-eyed hag in his path. She told him to get out of her way. He said that would leave him no room to pass except the sea-cliff below them. But she begged him to get out of her way. So he let her have the path, except where he clung by his toes. She struck at his big toe as she passed him by, to knock him off the path down the cliff. But he saw her in time and gave his hero's salmon-leap upward. Then he struck off the hag's head. She was Eis Enchenn, the bird-headed, mother of the three last warriors to die at his hands. It was to avenge their ruin that she lay in wait for him.

Soon Scáthach and her army went home to their own
country with the hostages that Aife gave them. Cúchulainn
waited there until his wounds were healed.

So Cúchulainn's training with Scáthach in the craft of
arms was done : what with the apple-feat — juggling nine
apples with never more than one in his palm; the thunder-
feat; the feats of the sword-edge and the sloped shield; the
feats of the javelin and rope; the body-feat; the feat of Cat
and the heroic salmon-leap; the pole-throw and the leap
over a poisoned stroke; the noble chariot-fighter's crouch;
the *gae bolga;* the spurt of speed; the feat of the chariot-
wheel thrown on high and the feat of the shield-rim; the
breath-feat, with gold apples blown up into the air; the
snapping mouth and the hero's scream; the stroke of precision; the stunning-shot and the cry-stroke; stepping on a
lance in flight and straightening erect on its point; the
sickle-chariot; and the trussing of a warrior on the points
of spears.

A message came for him to come back to his own
country. He bade them farewell. Scáthach spoke to him of
his future and his end. She chanted to him through the
imbas forasnai, the Light of Foresight. And this is what
she told him :

                    I salute you —
                            weary after triumph,
                    battle eager, ice hearted!
                            Go where you'll find some comfort still
                    what comfort comes with most speed
                            what speed with most urgency
                    alone no matter where you stand
                            dire danger ever at hand

alone and ringed by envy
>Cruachan's heroes you destroy
some heroes you protect
>others lie broken necked
your straight sword stabs behind you
>stained with Sétanta's own gore
red battle's distant roar
>bones broken by the spear
horned herds hemmed in
>the cruel club's hard edge
raw flesh battle's badge
>cattle stolen out of Breg
your country under bondage
>cattle straying on the ways
for five tear sodden days
>hardship and a long sigh
one against an army
>your own blood a red plague
splashed on many a smashed shield
>on weapons and women red eyed
the field of slaughter growing red
>on chopped flesh ravens feed
the crow scours the ploughed ground
>the savage kite shall be found
herds broken up in wrath
>great hosts driving the hordes
blood spilt in a great flood
>Cúchulainn's body wasted
there are bitter wounds to bear
>and warriors to slaughter
with your red stabbing spiked spear
>grief and sorrow where you roam
murderous on Murtheimne Plain
>playing at the stabbing game

now the crafty champion comes
         in rage against a broken wave
heroic in his mighty acts
         and harsh scream and cruel heart
let him come and women kill
         and Medb fight with Ailill
a bed of sickness lies in wait
         your breast full of fierce hate
hear the white horned bull roar
         against the brown bull of Cuailnge
when will he come and when force
         with sharp valour through the forest
arise versed in the bloody spike
         and long sweeping strong stroke
and twisting run and lone attack
         shake off weakness and neglect
arise once more and seize your arms
         seasoned in the crafts of war
proud striding raider pitiless
         for Ulster's land and virgin women
rise now in all your force
         with warlike cruel wounding shield
and strong shafted curved spear
         and straight sword dyed red
in dark gatherings of blood
         men in Alba will know your name
in the winter night pity your wail
         Aife and Uathach will pity
your sweet shape changing bright body
         stretched in sleep nobly broken
for three and thirty full years
         all your enemies are yours
you will keep for thirty years
         your sharp valour and your force

> I will not add another year
>> nor tell more of your career
> full of triumph and women's love
>> what matter how short.
> I salute you.'

Cúchulainn returned to Emain and told all his adventures. When he was rested he went on toward Forgall's ramparts in Luglochta Logo to find Emer. So strong was the guard around her that after a whole year there he hadn't reached her. Then he faced Forgall's fort in earnest. That day the sickle-chariot was harnessed for him and he drove it in a heavy course and did a thunder-feat that slew three hundred and nine men.

He reached Forgall's rampart and gave his salmon-leap across the three enclosures to the middle of the fort. In the inner enclosure he dealt three strokes at three groups of nine men. He killed eight men at each stroke and left one man standing in the middle of each group. They were Emer's three brothers, Scibar and Ibor and Cat. Forgall sprang away in flight from Cúchulainn out across the fort's rampart, but he fell and killed himself. Cúchulainn caught Emer and her foster-sister, and their weight in gold and silver, and leaped again with the two girls across the triple ramparts and hurried on with shrieks rising around them on every side. Scenmenn caught up with them but Cúchulainn killed him. Ath Scenmenn, Scenmenn's Ford, is the name of the place where he died. They went onward and came to Glondáth. There Cúchulainn killed one hundred men.

'That was a great deed,' Emer said, 'to kill one hundred armed angry men.'

'Glondáth, Ford of the Deed, will be its name forever,' Cúchulainn said.

Cúchulainn reached Crúfóit — which until then was called Rae Ban, the White Plain. He dealt the army great mortal blows there and streams of blood broke over the place on every side.

'You have made a hill of bloody sods today, Cúchulainn,' the girl said.

After this it was called Crúfóit, Sod of Blood.

The pursuers came up with them at Ath Imfóit on the Boann river. Emer got out of the chariot and Cúchulainn gave chase after the pursuers with the sods flying from his horses' hooves northward over the ford. Then he chased them northward and the sods flew southward from his horses' hooves over the ford. So it is called Ath Imfóit, from the sods flying this way and that.

Thus it came about that Cúchulainn killed one hundred men at every ford from Ath Scenmenn at Ailbine to the Boann at Breg, doing all he had told the girl he would do.

He reached Emain Macha safely at nightfall. Emer was brought into the Craebruad before Conchobor and the other Ulster chiefs and they made her welcome. But there was a sour sharp-tongued man there, Bricriu mac Carbad, and he said:

'Cúchulainn is going to find tonight's doings very hard. This woman he has brought here will have to sleep tonight with Conchobor — the first forcing of girls in Ulster is always his.'

Cúchulainn grew wild at this and trembled so hard that the cushion burst under him and the feathers flew around the house. He rushed out.

'This is very troublesome,' Cathbad said. 'The king can't refuse to do as Bricriu says. Yet Cúchulainn would destroy any man who slept with his wife.'

'Call Cúchulainn back,' Conchobor said, 'and we will try to cool his fever.'

Cúchulainn came in and Conchobor said:
'Go and bring me back all the herds about Sliab Fuait.'
Cúchulainn went off and gathered together all the swine and wild deer and every kind of wild flying creatures he could find at Sliab Fuait, and drove them in one flock on to the green at Emain. His anger was gone.

The men of Ulster argued the affair and decided that Emer should sleep that night in Conchobor's bed, but with Fergus and Cathbad in it as well to protect Cúchulainn's honour. They said the whole of Ulster would bless the couple if he accepted. He accepted, and so it was done. Conchobor paid Emer's dowry the next day, Cúchulainn was given his 'honour-price,' and he slept ever after with his wife. They never parted again until they died.

THE DEATH OF AIFE'S ONE SON

What caused Cúchulainn to kill his son?
It is soon told.

SEVEN years to the day after Cúchulainn left Aife, the boy came looking for his father. The men of Ulster were gathered at Tracht Esi, the Strand where the Mark is, when he came. They saw the boy coming toward them over the ocean in a little boat of bronze with gilt oars in his hands.

There was a pile of stones beside him in the boat. He put a stone in his sling and sent it humming at the sea birds, and stunned them without killing them. Then he let them escape into the air again. Then he did a feat with his jaws, between his hands, faster than the eye could follow, tuning his voice to bring them down a second time. Then he roused them again.

'Well,' Conchobor said, 'I pity the country that boy is heading for. I don't know what island he comes from, but their grown men can grind us into dust if one of their young boys can do that. Someone go out to meet him. Don't let him ashore.'

'Who ought to meet him?'

'Who but Condere mac Echach?' Conchobor said.

'And why Condere?' they all asked.

'Clearly,' Conchobor said, 'where there is a need for good sense and eloquence, Condere is the right person.'

'I'll go and meet him,' Condere said.

Condere went up to the boy just as he reached the strand.

'You have come far enough, young man,' Condere said, 'until we find out where you come from and who your people are.'

'I'll give my name to no man,' the boy said, 'and I'll make way for no man.'

'You can't land', Condere said, 'unless you give your name.'

'I am going where I am going,' the boy said.

The boy moved to pass him, but Condere said:

> 'Heed me my son.
> Mighty are your acts
>> manly your blood
> you have the pride
>> of an Ulster warrior
> Conchobor would protect you
>> but you bare your jaws
> and dare us with your little spears
>> and annoy our warriors
> you have come to Conchobor
>> let him grant you protection.
> Listen, pay heed.

Come to Conchobor
    Nes's swift son
to Sencha mac Ailella
    full of victories
to Fintan's son Cethern
    of the crimson blade
the fire that burns battalions
    to the poet Amargin
to Cúscraid of the huge hosts
    come into the care
of Conall Cernach
    above story or song
or the shouts of heroes
    gathered together
Blai Briuga would dislike it
    if you pushed past him
or any warrior
    however fine
the insult would hurt him
    come let it be said
that Condere himself
    arose and approached
the warlike boy
    and held him back.

'I have sworn to oppose you, a beardless, unfledged boy,' Condere said, 'if you won't heed the men of Ulster.'

'You have come and spoken well,' the boy said, 'so I will answer you:

'I tuned my voice:
from little jaws
    a straight shot sped
with my little spears
    flung from afar

> I gathered together
>   a lovely bird flock
> no need of my hero's
>   salmon leap
> by such brave acts
>   I have sworn no man
> will stand in my way
>   go back to Ulster
> and say I'll fight them
>   singly or together.

'Turn aside,' the boy said, 'for even though you had the strength of a hundred men you couldn't hold me back.'

'Very well,' Condere said, 'someone else can try.'

Condere went back to the men of Ulster and told them.

'No one makes little of Ulster's honour while I live,' Conall Cernach said. 'I won't permit it.'

He went out to meet the boy.

'Those were pretty games, boy,' Conall said.

'They'll work just as nicely on you,' the boy said.

He set a stone in his sling and sent it in a stunning-shot into the sky. The roar of its thunder as it rose reached Conall and knocked him headlong. Before he could rise the boy had the shield-strap tied around Conall's arms.

'Send out someone else!' Conall said, but the whole army was put to shame.

Then Cúchulainn advanced on the boy, performing his feats as he came. Forgall's daughter Emer had her arm round his neck. She said:

> 'Don't go down!
> It is your own son there
>   don't murder your son
> the wild and well born
>   son let him be

            is it good or wise
                        for you to fall
            on your marvellous son
                        of the mighty acts
            remember Scáthach's
                        strict warning and turn
            from this flesh agony
                        this twig from your tree
            if Connla has dared us
                        he has justified it.

            Turn back, hear me!
            My restraint is reason
                        Cúchulainn hear it
            we know his name
                        if he is really Connla
            the boy is Aife's
                        one son.'

Then Cúchulainn said:

            'Be quiet, wife.
                        It isn't a woman
            that I need now
                        to hold me back
            in the face of these feats
                        and shining triumph
            I want no woman's
                        help with my work
            victorious deeds
                        are what we need
            to fill the eyes
                        of a great king
            the blood of Connla's
                        body will flush

>   my skin with power
>   little spear so fine
>   to be finely sucked
>   by my own spears!

'No matter who he is, wife,' Cúchulainn said, 'I must kill him for the honour of Ulster.'

So he went down to meet him.

'Those were pretty games, boy,' he said.

'Prettier than the games I'm finding here,' the young boy said. 'Two of you have come down here and still I haven't named myself.'

'Maybe you were meant to meet me,' Cúchulainn said. 'Name yourself, or you die.'

'So be it!' the boy said.

The boy set upon him and they struck at one another. The boy cut him bald-headed with his sword, in the stroke of precision.

'The joking has come to a head!' Cúchulainn said. 'Now we'll wrestle.'

'I can't reach up to your belt,' the boy said.

He climbed up onto two standing stones. Without moving a foot he trust Cúchulainn three times between the two stones. His feet sank in the stone up to the ankle. The marks of his feet are there still, which is why the people of Ulster call it Tráig, or Tracht. Esi, the Strand of the Mark.

They went down into the sea to drown each other, and the boy submerged him twice. Then Cúchulainn turned and played the boy foul in the water with the *gae bolga*, that Scáthach had taught to no one but him. He sent it speeding over the water at him and brought his bowels down around his feet.

'There is something Scáthach didn't teach me,' the boy

said. 'You have wounded me woefully.'

'I have,' Cúchulainn said.

He took the boy in his arms and carried him away from the place and brought him and laid him down before the people of Ulster.

'My son, men of Ulster,' he said. 'Here you are.'

'Alas, alas!' said all Ulster.

'It is the truth,' the boy said. 'If only I had five years among you I would slaughter the warriors of the world for you. You would rule as far as Rome. But since it is like this, point me out the famous men around me. I would like to salute them.'

He put his arms round the neck of each man in turn, and saluted his father, and then died. Then a loud lament was uttered for him. His grave was made and the gravestone set. For the space of three days and nights no calf in Ulster was let go to its cow on account of his death.

THE QUARREL OF THE TWO PIG-KEEPERS, AND HOW THE BULLS WERE BEGOTTEN

What caused the two pig-keepers to quarrel?
It is soon told.

THERE was bad blood between Ochall Ochne, the king of the *síd* in Connacht, and Bodb, king of the Munster *síd*. (Bodb's *síd* is the 'Síd ar Femen,' the *síd* on Femen Plain; Ochall's is the *síd* at Cruachan.) They had two pig-keepers, called Friuch, after a boar's bristle, and Rucht, after its grunt. Friuch was Bodb's pig-keeper, Rucht was Ochall's, and they were good friends. They were both practised in the pagan arts and could form themselves into any shape, like Mongán mac Fiachna.

The two pig-keepers were on such good terms that the one from the north would bring his pigs down with him when there was a mast of oak and beech nuts in Munster. If the mast fell in the north the pig-keeper from the south would travel northward.

There were some who tried to make trouble between them. People in Connacht said their pig-keeper had the greater power, while others in Munster said it was theirs who had greater power. A great mast fell in Munster one

year, and the pig-keeper from the north came southward
with his pigs. His friend made him welcome.

'Is it you?' he said. 'They are trying to cause trouble
between us. Men here say your power is greater than
mine.'

'It is no less, anyway,' Ochall's pig-keeper said.

'That's something we can test,' Bodb's pig-keeper said.
'I'll cast a spell over your pigs. Even though they eat this
mast they won't grow fat, while mine will.'

And that is what happened. Ochall's pig-keeper had to
bring his pigs away with him so lean and wretched that
they hardly reached home. Everybody laughed at him as
he entered his country.

'It was a bad day you set out,' they said. 'Your friend
has greater power than you.'

'It proves nothing,' he said. 'We'll have mast here in
our own turn and I'll play the same trick on him.'

This also happened. Bodb's pig-keeper came northward
the same time next year into the country of Connacht,
bringing his lean pigs with him, and Ochall's pig-keeper
did the same to them, and they withered. Everybody said
then that they had equal power. Bodb's pig-keeper came
back from the north with his lean pigs, and Bodb dismissed him from pig-keeping. His friend in the north
was also dismissed.

After this they spent two full years in the shape of
birds of prey, the first year at the fort of Cruachan, in
north Connacht, and the second at the *síd* on Femen
Plain. One day the men of Munster had collected together
at this place.

'Those birds are making a terrible babble over there,'
they said. 'They have been quarrelling and behaving like
this for a full year now.'

As they were talking they saw Fuidell mac Fiadmire,

Ochall's steward, coming toward them up the hill and they made him welcome.

'Those birds are making a great babble over there,' he said. 'You would swear they were the same two birds we had back north last year. They kept this up for a whole year.'

Then they saw the two birds of prey turn suddenly into human shape and become the two pig-keepers. They made them welcome.

'You can spare your welcome,' Bodb's pig-keeper said. 'We bring you only war-wailing and a fullness of friends' corpses.'

'What have you been doing?' Bodb said.

'Nothing good,' he said. 'From the day we left until today we spent two full years together in the shape of birds. You saw what we did over there. A whole year went like that at Cruachan and a year at the *síd* on Femen Plain so that all men, north and south, have seen our power. Now we are going to take the shape of water creatures and live two years under the sea.'

They left and each went his own way. One entered the Sinann river, the other the river Siuir, and they spent two full years under water. One year they were seen devouring each other in the Siuir, the next in the Sinann.

Next they turned into two stags, and each gathered up the other's herd of young deer and made a shambles of his dwelling place.

Then they became two warriors, gashing each other.

Then two phantoms, terrifying each other.

Then two dragons, pouring down snow on each other's land.

They dropped down then out of the air, and became two maggots. One of them got into the spring of the river Cronn in Cuailnge, where a cow belonging to Dáire mac

Fiachna drank it up. The other got into the well-spring
Garad in Connacht, where a cow belonging to Medb and
Ailill drank it. From them, in this way, sprang the two
bulls, Finnbennach, the white-horned, of Ai Plain, and
Dub, the dark bull of Cuailnge.

Rucht and Friuch were their names when they were
pig-keepers; Ingen and Eitte, Talon and Wing, when they
were two birds of prey; Bled and Blod, Whale and
Seabeast, when they were two undersea creatures; Rinn
and Faebur, Point and Edge, when they were two warriors;
Scáth and Sciath, Shadow and Shield, when they were two
phantoms; and Cruinniuc and Tuinniuc when they were
two maggots. Finnbennach Ai, the White, and Donn
Cuailnge, the Brown, were their names when they were
two bulls.

This was the Brown Bull of Cuailnge —
    dark brown dire haughty with young health
    horrific overwhelming ferocious
    full of craft
    furious fiery flanks narrow
    brave brutal thick breasted
    curly browed head cocked high
    growling and eyes glaring
    tough maned neck thick and strong
    snorting mighty in muzzle and eye
    with a true bull's brow
    and a wave's charge
    and a royal wrath
    and the rush of a bear
    and a beast's rage
    and a bandit's stab
    and a lion's fury.

Thirty grown boys could take
their place from rump to nape
— a hero to his herd at morning
foolhardy at the herd's head
to his cows the beloved
to husbandmen a prop
the father of great beasts
overlooks the ox of the earth.

A white head and white feet
    had the Bull Finnbennach
    and a red body the colour of blood
    as if bathed in blood
    or dyed in the red bog
    or pounded in purple
    with his blank paps
    under breast and back
    and his heavy mane and great hoofs
    the beloved of the cows of Ai
    with ponderous tail
    and a stallion's breast
    and a cow's eye apple
    and a salmon's snout
    and hinder haunch
    he romps in rut
    born to bear victory
    bellowing in greatness
    idol of the ox herd
    the prime demon Finnbennach.

THE TAIN

I THE PILLOW TALK

ONCE when the royal bed was laid out for Ailill and Medb in Cruachan fort in Connacht, they had this talk on the pillows:

'It is true what they say, love,' Ailill said, 'it is well for the wife of a wealthy man.'

'True enough,' the woman said. 'What put that in your mind?'

'It struck me,' Ailill said, 'how much better off you are today than the day I married you.'

'I was well enough off without you,' Medb said.

'Then your wealth was something I didn't know or hear much about,' Ailill said. 'Except for your woman's things, and the neighbouring enemies making off with loot and plunder.'

'Not at all,' Medb said, 'but with the high king of Ireland for my father — Eochaid Feidlech the steadfast, the son of Finn, the son of Finnoman, the son of Finnen, the son of Finngoll, the son of Roth, the son of Rigéon, the son of Blathacht, the son of Beothacht, the son of Enna Agnech, the son of Aengus Turbech. He had six daughters: Derbriu, Ethne, Ele, Clothru, Muguin, and myself Medb, the highest and haughtiest of them. I outdid them in grace and giving and battle and warlike

combat. I had fifteen hundred soldiers in my royal pay,
all exiles' sons, and the same number of freeborn native
men, and for every paid soldier I had ten more men, and
nine more, and eight, and seven, and six, and five, and
four, and three, and two, and one. And that was only our
ordinary household.

'My father gave me a whole province of Ireland, this
province ruled from Cruachan, which is why I am called
"Medb of Cruachan." And they came from Finn the king
of Leinster, Rus Ruad's son, to woo me, and from Coirpre
Niafer the king of Temair, another of Rus Ruad's sons.
They came from Conchobor, king of Ulster, son of
Fachtna, and they came from Eochaid Bec, and I wouldn't
go. For I asked a harder wedding gift than any woman
ever asked before from a man in Ireland — the absence
of meanness and jealousy and fear.

'If I married a mean man our union would be wrong,
because I'm so full of grace and giving. It would be an
insult if I were more generous than my husband, but not
if the two of us were equal in this. If my husband was a
timid man our union would be just as wrong because I
thrive, myself, on all kinds of trouble. It is an insult for
a wife to be more spirited than her husband, but not if the
two are equally spirited. If I married a jealous man that
would be wrong, too: I never had one man without
another waiting in his shadow. So I got the kind of man
I wanted: Rus Ruad's other son — yourself, Ailill, from
Leinster. You aren't greedy or jealous or sluggish. When
we were promised, I brought you the best wedding gift a
bride can bring: apparel enough for a dozen men, a
chariot worth thrice seven bondmaids, the width of your
face of red gold and the weight of your left arm of light
gold. So, if anyone causes you shame or upset or trouble,
the right to compensation is mine,' Medb said, 'for you're

a kept man.'

'By no means,' Ailill said, 'but with two kings for my brothers, Coirpre in Temair and Finn over Leinster. I let them rule because they were older, not because they are better than I am in grace or giving. I never heard, in all Ireland, of a province run by a woman except this one, which is why I came and took the kingship here, in succession to my mother Mata Muiresc, Mágach's daughter. Who better for my queen than you, a daughter of the high king of Ireland?'

'It still remains,' Medb said, 'that my fortune is greater than yours.'

'You amaze me,' Ailill said. 'No one has more property or jewels or precious things than I have, and I know it.'

Then the lowliest of their possessions were brought out, to see who had more property and jewels and precious things: their buckets and tubs and iron pots, jugs and wash-pails and vessels with handles. Then their finger-rings, bracelets, thumb-rings and gold treasures were brought out, and their cloth of purple, blue, black, green and yellow, plain grey and many-coloured, yellow-brown, checked and striped. Their herds of sheep were taken in off the fields and meadows and plains. They were measured and matched, and found to be the same in numbers and size. Even the great ram leading Medb's sheep, the worth of one bondmaid by himself, had a ram to match him leading Ailill's sheep.

From pasture and paddock their teams and herds of horses were brought in. For the finest stallion in Medb's stud, worth one bondmaid by himself, Ailill had a stallion to match. Their vast herds of pigs were taken in from the woods and gullies and waste places. They were measured and matched and noted, and Medb had one fine boar, but Ailill had another. Then their droves and free-wandering herds of cattle were brought in from the woods and wastes of the province. These were matched and measured and noted also, and found to be the same in number and size. But there was one great bull in Ailill's herd, that had been a calf of one of Medb's cows — Finnbennach was his name, the White Horned — and Finnbennach, refusing to be led by a woman, had gone over to the king's herd. Medb couldn't find in her herd the equal of this bull, and her spirits dropped as though she hadn't a single penny.

Medb had the messenger Mac Roth called, and she told him to see where the match of the bull might be found, in any province in Ireland.

'I know where to find such a bull and better,' Mac Roth said: 'in the province of Ulster, in the territory of Cuailnge, in Dáire mac Fiachna's house. Donn Cuailnge is the bull's name, the Brown Bull of Cuailnge.'

'Go there, Mac Roth,' Medb said. 'Ask Dáire to lend me Donn Cuailnge for a year. At the end of the year he can have fifty yearling heifers in payment for the loan, and the Brown Bull of Cuailnge back. And you can offer him this too, Mac Roth, if the people of the country think badly of losing their fine jewel, the Donn Cuailnge: if Dáire himself comes with the bull I'll give him a portion of the fine Plain of Ai equal to his own lands, and a chariot worth thrice seven bondmaids, and my own friendly thighs on top of that.'

Messengers set out to Dáire mac Fiachna's house: there were nine of them with Mac Roth. Mac Roth was soon made welcome in Dáire's house, as befitted Ireland's chief messenger. Dáire asked him what brought him on his journey, and the chief messenger told him why he came, and about the squabble between Medb and Ailill.

'So I am here to ask for the loan of the Donn Cuailnge, to match against Finnbennach,' he said. 'And you'll get

fifty yearling heifers back in payment for the loan, with
the Donn Cuailnge himself and more besides. If you come
with the bull yourself you'll get a portion of the fine
Plain of Ai equal to your own lands, and a chariot worth
thrice seven bondmaids, and Medb's friendly thighs on
top of it all.'

Dáire was delighted, and jumped for joy till the seams
of his cushion burst under him, and he cried:

'True as my soul! I don't care what the Ulstermen
think, I'll take my treasure, the Donn Cuailnge, to Ailill
and Medb in the land of Connacht.'

Mac Roth was pleased at mac Fiachna's decision.

Then they were looked after, and rushes and fresh
straw were settled under them. They were given the
best of good food and kept supplied with the festive fare
until they grew drunk and noisy.

Two of the messengers were talking. One of them said:

'There's no doubt, the man of the house here is a good
man.'

'A good man certainly,' the other said.

'Is there a better man in Ulster?' the first messenger
said.

'There is, certainly,' the second messenger said. 'His
leader Conchobor is a better man. If the whole of Ulster
gave in to him, it would be no shame for them. Anyway,
it was good of him to give us the Donn Cuailnge. It would
have taken four strong provinces of Ireland to carry it
off from Ulster otherwise.'

A third man joined the talk.

'What are you arguing about?' he said.

'This messenger here said, "The man of the house here
is a good man." "A good man certainly," the other said.
"Is there a better man in Ulster?" the first messenger said.
"There is, certainly," the second messenger said. "His

leader Conchobor is a better man. If the whole of Ulster
gave in to him, it would be no shame for them. But it was
good of him to give us what the four strong provinces of
Ireland would be needed to take from Ulster." '

'I'd as soon see the mouth that said that spout blood!
We would have taken it anyway, with or without his
leave.'

At that moment the man in charge of Dáire mac
Fiachna's household came into the hut, with a man carrying drink and another man with food, and heard what
they were saying. He was seized with fury, and put down
their food and drink, saying neither 'Eat' nor 'Don't eat.'
He went back straight to Dáire mac Fiachna's hut and
said:

'Did you give our famous treasure, the Donn Cuailnge,
to Medb's messengers?'

'Yes I did,' Dáire said.

'That was not a kingly thing to do. What they said is
true: if you hadn't given him up freely the hosts of Ailill
and Medb, and the cunning of Fergus mac Roich, would
have had him without your leave.'

'By the gods I worship, nothing leaves here unless I
choose to let it!'

They waited until morning. The messengers got up
early the next day and went to Dáire's hut.

'Tell us, sir, where to find the Donn Cuailnge.'

'I will not,' Dáire said. 'And only it isn't my habit to
murder messengers or travellers or any other wayfarers,
not one of you would leave here alive.'

'Why is this?' Mac Roth said.

'For a good reason,' Dáire said. 'You said if I didn't
give willingly, the hosts of Ailill and Medb, and Fergus's
cunning, would make me give.'

'Indeed,' Mac Roth said, 'what messengers say into

your food and drink hardly deserves your notice. You
can't blame Ailill and Medb.'

'Still, I won't give up my bull this time, Mac Roth, as
long as I can help it.'

So the messengers set off again and came to Cruachan,
the stronghold of Connacht. Medb asked them for the
news and Mac Roth said Dáire wouldn't give up his bull.

'Why not?' Medb said.

Mac Roth told what had happened.

'We needn't polish the knobs and knots in this, Mac
Roth,' Medb said. 'It was well known it would be taken by
force if it wasn't given freely. And taken it will be.'

II   THE TAIN BO CUAILNGE BEGINS

AILILL and Medb assembled a great army in
Connacht, and they sent word also to
the other three provinces. Ailill sent out messengers as
well to his brothers, the rest of Mágach's seven sons.
Besides Ailill there were Anluan, Mugcorb, Cet, En,
Bascall and Dóchae, and each of them had a troop of
three thousand. And he sent to Conchobor's son, Cormac
Connlongas, the leader of the Ulster exiles, and his troop
of three thousand who were living in Connacht. Soon they
all came to Cruachan Ai.

Cormac, marching to Cruachan, had three companies.
The first company wore speckled cloaks wrapped around
them. Their hair was clipped. Tunics covered them to the
knee. They carried full-length shields and each man had
a broad grey stabbing-spear on a slender shaft. The second
company wore dark-grey cloaks around them and red-
embroidered tunics that reached to their calves. Their hair
was drawn back on their heads and they carried bright

shields before them and five-pronged spears in their hands.

'I don't see Cormac yet,' Medb said.

Then the third troop came up. They wore purple cloaks and red-embroidered hooded tunics reaching to their feet. Their hair was trimmed to the shoulder. They carried curved scallop-edged shields, and a spear like a palace pillar in each man's hand.

'Now I see him,' Medb said.

Four of the provinces of Ireland gathered there at Cruachan Ai. Their sages and druids delayed them there for a fortnight waiting for a sign. The day they finally set out Medb said to her charioteer:

'Everyone leaving a lover or a friend today will curse me,' she said. 'This army is gathered for me.'

'Wait a minute,' the charioteer said, 'until I turn the chariot around to the right, with the sun, to draw down the power of the sign for our safe return.'

He turned the chariot round and made to set off. But they saw a young grown girl in front of them. She had yellow hair. She wore a speckled cloak fastened around her with a gold pin, a red-embroidered hooded tunic and sandals with gold clasps. Her brow was broad, her jaw narrow, her two eyebrows pitch black, with delicate dark lashes casting shadows half way down her cheeks. You would think her lips were inset with Parthian scarlet. Her teeth were like an array of jewels between the lips. She had hair in three tresses: two wound upward on her head and the third hanging down her back, brushing her calves. She held a light gold weaving-rod in her hand, with gold inlay. Her eyes had triple irises. Two black horses drew her chariot, and she was armed.

'What is your name?' Medb said to the girl.

'I am Fedelm, and I am a woman poet of Connacht.'

'Where have you come from?' Medb said.

'From learning verse and vision in Alba,' the girl said.

'Have you the *imbas forasnai*, the Light of Foresight?' Medb said.

'Yes I have,' the girl said.

'Then look for me and see what will become of my army.'

So the girl looked.

Medb said, 'Fedelm, prophetess; how seest thou the host?'

Fedelm said in reply:

'I see it crimson, I see it red.'

'It can't be true,' Medb said. 'Conchobor is suffering his pangs in Emain with all the rest of the Ulster warriors. My messengers have come from there and told me. Fedelm, prophetess; how seest thou our host?'

'I see it crimson, I see it red,' the girl said.

'That is false,' Medb said. 'Celtchar mac Uthidir is still in Dún Lethglaise with a third of Ulster's forces, and Fergus son of Roech mac Echdach and his troop of three thousand are here with us in exile. Fedelm, prophetess; how seest thou our host?' Medb said.

'I see it crimson, I see it red,' the girl said.

'It doesn't matter,' Medb said. 'Wrath and rage and red wounds are common when armies and large forces gather. So look once more and tell us the truth. Fedelm, prophetess; how seest thou our host?'

'I see it crimson, I see it red,' the girl said.

> 'I see a battle: a blond man
> with much blood about his belt
> and a hero-halo round his head.
> His brow is full of victories.

Seven hard heroic jewels
are set in the iris of his eye.
His jaws are settled in a snarl.
He wears a looped, red tunic.

A noble countenance I see,
working effect on womenfolk;
a young man of sweet colouring;
a form dragonish in the fray.

His great valour brings to mind
Cúchulainn of Murtheimne,
the hound of Culann, full of fame.
Who he is I cannot tell
but I see, now, the whole host
coloured crimson by his hand.

A giant on the plain I see,
doing battle with the host,
holding in each of his two hands
four short quick swords.

I see him hurling against that host
two *gae bolga* and a spear
and an ivory-hilted sword,
each weapon to its separate task.

He towers on the battlefield
in breastplate and red cloak.
Across the sinister chariot-wheel
the Warped Man deals death
— that fair form I first beheld
melted to a mis-shape.

I see him moving to the fray :
take warning, watch him well,
Cúchulainn, Sualdam's son !
Now I see him in pursuit.

Whole hosts he will destroy,
making dense massacre.
In thousands you will yield your heads.
I am Fedelm. I hide nothing.

The blood starts from warrior's wounds
— total ruin — at his touch :
your warriors dead, the warriors
of Deda mac Sin prowling loose;
torn corpses, women wailing,
because of him — the Forge-Hound.'

The Monday after Samain they set out. This is the way
they went, southeast from Cruachan Ai :
   through Muicc Cruinb,
   through Terloch Teóra Crích, the marshy lake bed
   where three territories meet,
   by Tuaim Móna, the peat ridge,
   through Cúil Silinne, where Carrcin Lake is now — it
   was named after Silenn, daughter of Madchar,·
   by Fid and Bolga, woods and hills,
   through Coltain, and across the Sinann river,
   through Glúne Gabair,
   over Trego Plain, of the spears,
   through Tethba, North and South,
   through Tiarthechta,
   through Ord, 'the hammer,'
   through Sláis southward,
   by the river Indiuind, 'the anvil,'

through Carn,
through Ochtrach, 'the dung heap,'
through Midi, the land of Meath,
through Finnglassa Assail, of the clear streams,
by the river Deilt,
through Delind,
through Sailig,
through Slaibre of the herds,
through Slechta, where they hewed their way,
through Cúil Sibrille,
southward by Ochaine hill,
northward by Uatu,
by the river Dub,
southward through Comur,
through Tromma,
through Othromma eastward,
through Sláni and its pasture Gortsláni,
southward by Druim Licce, 'the flagstone ridge,'
by Ath Gabla, the ford of the forked branch,
through Ard Achad, the high field,
northward by Féraind,
by Finnabair,
through Assi southward,
by the ridge Druim Sálfinn,
by the ridge Druim Cain, on the Midluachair road,
by mac Dega's ridge,
by Eódond Mór and Eódond Bec, the great dark yew-
tree and the lesser,
by Méthe Tog and Méthe nEoin, 'squirrel neck' and
'bird neck,'
by the ridge Druim Cáemtechta,
through Scúaip and Imscúaip,
through Cenn Ferna,
through Baile and Aile,

through Báil Scena and Dáil Scena,
through Fertse,
by the wooded promontory Ros Lochad,
through Sale,
through Lochmach, or Muid Loga,
through Anmag, the noble plain,
by Dinn height,
by the river Deilt,
by the river Dubglais,
through Fid Mór, or Fid Mórthruaille, the Wood of the
Great Scabbard,
to the river Colptha
and to the river Cronn in Cuailnge.

These are the places they were to pass on their way to
Finnabair in Cuailnge. It was from Finnabair that the
armies of Ireland later split up across the province to look
for the bull.

III   THE ARMY ENCOUNTERS CUCHULAINN

ON the first stage of their march they went from
Cruachan to Cúil Silinne, at Carrcin Lake.
Medb told her charioteer to yoke up her nine chariots
ready to made a circuit of the camp, to see who was slow
and who eager on the march.

Meanwhile Ailill's tent was pitched and his things
settled, the beds and coverlets. Next to Ailill came Fergus
mac Roich in his tent; next to Fergus, Cormac Connlongas,
Conchobor's son; next to him, Conall Cernach; and next
to him, Fiacha mac Fir Febe, the son of Conchobor's
daughter. Medb was to settle the other side of Ailill; next
to her their daughter, Finnabair; and next to her, Flidais.   Flidash
Not to speak of menservants and attendants.

Medb came back from inspecting the armies and said

Maive

it would be foolish to go on if they let the troop of three thousand Galeóin, from north Leinster, come with them.

'What fault have you found with them?' Ailill said.

'I find no fault with them,' Medb said. 'They are fine soldiers. While the others were making a space for their camp they had roofed theirs and were making their meal. While the others were eating they had finished their meal and had their harpers playing. So it would be foolish to take them,' Medb said. 'They would get all the credit for our army's triumph.'

'But they are fighting on our side,' Ailill said.

'They can't come,' Medb said.

'Let them stay, then,' Ailill said.

'No, they can't stay either,' Medb said. 'They would only come and seize our lands when we are gone.'

'Well, what are we going to do with them,' Ailill said, 'if they can neither stay nor come?'

'Kill them,' Medb said.

'That is a woman's thinking and no mistake!' Ailill said. 'A wicked thing to say.'

'These men are our friends,' Fergus said, for the Ulster exiles. 'You will take this evil advice over our dead bodies.'

'We might do that,' Medb said. 'I have my own following of twice three thousand here. There are my sons too, the seven Maine, with their seven troops of three thousand — may they always have luck. There is Maine Máthramail the Motherlike, Maine Athramail the Fatherlike, Maine Mórgor the strongly dutiful, Maine Míngor the sweetly dutiful, Maine Móepirt who is above description (some call him Maine Milscothach of the honeyed speech), Maine Andoe the swift, and Maine Cotagaib Uli — the Maine with all the qualities, who took the likeness of his mother and father, and the dignity of us both.'

'That is not the whole story,' Fergus said. 'We have

seven Munster kings on our side, each with a troop of
three thousand. Here and now, in this camp, I could bring
those seven troops of three thousand into battle against
you, with my own three thousand and the Galeóin troop.
But we don't need that,' Fergus said. 'We can arrange
these warriors in the army so that they won't stand out
too much. There are seventeen troops of us, of three
thousand each,' Fergus said; 'that is the full number of
our camp, not counting the general rabble or the young
or the women — each king has his queen travelling with
Medb. The Galeóin troop is the eighteenth troop here.
We can scatter them out among the whole army.'

'I don't mind,' Medb said, 'as long as they break up
their present order.'

So that is what they did: the Galeóin were scattered
amongst the army.

Next morning they set out toward Móin Coltna, the
moor near Coltain. They found eight score of wild deer
there in one herd, and encircled and slaughtered them.
Wherever there was one of the Galeóin it was he who
got the deer, except for five that were got by the rest of
the army.

They came to Trego Plain and broke their march there
and got their meal ready. They say it is here that Dubthach
chanted:

> 'Take note now, listen well
> to my vision of this war.
> A dark march lies ahead
> toward Ailill's wife's White Horn.
>
> One man, worth a whole host,
> comes to guard Murtheimne's herds.
> Two pig-keepers were friends once —
> now crows will drink a cruel milk.

The river Cronn will rise, all clay,
and bar the way to Murtheimne
until that warrior's work is done
at Mount Ochaine to the north.

"Quickly," Ailill says to Cormac,
"hurry to your son's side!"
Cattle calm upon the plains —
the hard raiders herding men.

Then a battle, in due time,
with Medb and one third of the host —
man's meat everywhere
that the Warped Man can reach.'

Immediately the war-spirit Nemain assailed them. They
had no peace that night, with their sleep broken by
Dubthach's brute outcry. Groups of them started up,
and many of the army remained troubled until Medb
came and calmed them. They went on then and spent the
night in Gránaird in North Tethba.

Fergus sent a warning from there to the men of Ulster,
because of old friendship. They were still prostrate in
their pangs, all but Cúchulainn and his father Sualdam.
When Fergus's warning came Cúchulainn and his father
went out as far as Iraird Cuillenn and set up watch for
the armies there.

'I feel the presence of the armies tonight,' Cúchulainn
said to his father. 'You must go and warn Ulster. I have
promised to spend the night with Fedelm Noichride.'
(Though some say his meeting was with her bondmaid,
who was set aside for Cúchulainn's use.) Before he left
he made a spancel-hoop of challenge and cut an ogam
message into the peg fastening it, and left it there for
them on top of a standing stone.

Fergus was given the head of the army, out in front of
the troops. He made a great detour southward to give
Ulster time to gather an army together — he did this out
of old friendship. But Ailill and Medb noticed it, and
Medb said:

> 'Fergus, there is something wrong.
> What kind of road is this we're taking?
> — straying to the south or north,
> crossing every kind of land.
>
> Ailill and his army
> begin to think of treachery.
> Or have you not yet set your mind
> to leading us upon our way?
>
> If old friendship is the cause
> give up your first place on the march.
> Perhaps another can be found
> to take us on our proper way.'

Fergus answered:

> 'Medb, what is troubling you?
> There's no treachery in this.
> The land where I am taking you
> — remember it is Ulster.
>
> I take these turnings as they come
> not to bring the host to harm
> but to miss the mighty man
> who protects Murtheimne Plain.
>
> Do you think I don't know
> every winding way I take?
> I think ahead, trying to miss
> Cúchulainn son of Sualdam.'

Cu huellen

Then they came to Iraird Cuillenn.

Err and Innel and their two charioteers Foich and
Fochlam (these were the four sons of Urard mac Anchinne)
were out in front of the army, keeping their rugs and
cloaks and brooches from being soiled by the dusts of the
multitude. They found the spancel-hoop thrown there by
Cúchulainn and saw the marks of how his horses had
grazed. Sualdam's two horses had bitten the grass, roots
and all, out of the earth, while Cúchulainn's horses had
licked up the very clay as well, down to the stones beneath
the grass.

They sat and waited for the armies to come up, while
their musicians played for them.

They gave the spancel-hoop to Fergus mac Roich, and
he read the ogam cut into the hoop.

When Medb came up she said:

'Why are you waiting here?'

'We are waiting because of this spancel-hoop,' Fergus
said. 'There is an ogam message on the peg. It says:
"Come no further, unless you have a man who can make
a hoop like this with one hand out of one piece. I exclude
my friend Fergus." It is clear Cúchulainn did this,' Fergus
said. 'It was his horses that grazed the plain.'

Fergus gave it into the druids' hands and chanted:

> 'This hoop: what does it mean to us?
> What is the riddle of the hoop?
> How many men put it here?
> A small number? A multitude?
>
> Will it bring the host to harm
> if they pass it on their way?
> Druids, discover if you can
> the reasons it was left here.'

The druids answered:
> 'It was a great champion made it
> and left it as a trap for men,
> an angry barrier against kings
> — one man, single-handed.

> The royal host must come no further,
> according to the rule of war,
> unless you have a man among you
> who can do what he has done.
> This is the reason, and no other,
> why the spancel-hoop was left.'

Then Fergus told them:
'If you ignore this challenge and pass by, the fury of the man who cut that ogam will reach you even if you are under protection, or locked in your homes. Unless someone can match this hoop of challenge he will kill one of you before morning.'

'We have certainly no wish to see one of our men killed so soon,' Ailill said. 'If we go through the neck of that great forest there south of us, Fid Dúin, we needn't pass here.'

Then the men of the armies cut down the forest before the chariots. The place is now called Slechta, the Hewn Place.

It is told in other books that it was after they had reached Fid Dúin, the forest fortress, that they saw the chariot with the beautiful young girl — that it was here the story of the prophetess Fedelm, already given above, took place, and that the forest was cut down after a certain answer she gave to Medb. ("Look for me, and see what will became of my army," Medb had said. "It is too hard," the girl said: "I can't see them properly in this

Fedelm

forest." "Then it will be made ploughed land," Medb
said: "we will cut down the forest.") And that thus the
place is called Slechta.

It is here that the Partraigi dwell.

They passed the night in Cúil Sibrille — Cenánnos,
as it is now called. A great snow fell on them, over
the men's belts and the chariot wheels. They could
get no food ready, and rose early the next day, after
passing a hard night in the snow.

But it wasn't so early when Cúchulainn got up from his
woman, and it was later still when he had scrubbed and
scoured himself and found the track of the army.

'I wish we hadn't gone there,' Cúchulainn said, 'and
betrayed Ulster. We let an army through and gave no
warning. Reckon up the army's tracks for us,' Cúchulainn
said to Laeg, 'until we see how many of them there are.'

Laeg did so, and said to Cúchulainn:

'This is confusing. I can't reckon it.'

'It wouldn't confuse me if I went up there.' Cúchulainn
said.

'Get into the chariot then,' Laeg said.

Cúchulainn got into the chariot and tried for a long
time to reckon up the army from their tracks.

'Even you don't find it easy,' Laeg said.

'Still it is easier for me than you,' Cúchulainn said,
'with my three talents of sight and intellect and reckoning.
I have made up a count now,' he said. 'There are eighteen
troops of three thousand here, as I count them, but the
eighteenth troop of three thousand has been divided out
among the whole army. That is what is mixing up the
count, the three thousand Galeóin.'

Cúchulainn went around the armies until he reached

Gow blc

Ath Gabla. There he cut out a tree-fork with a single stroke of his sword and stuck it in the middle of the stream, so that a chariot would have no room to pass it on either side. (It is from this that the name Ath Gabla comes, the ford of the forked branch.) The warriors Err and Innel, and their two charioteers Foich and Fochlam, came upon him. He cut off their four heads and tossed them onto the four points of the tree-fork.

The horses of the four men went back toward the army with their coverings all crimson. Everyone thought there was a battle-force waiting for them at the ford. A troop of them went to inspect the ford, but they saw nothing there except the track of a single chariot, and the fork with the four heads and the words in ogam cut into its side. Then the whole army came up.

'Do these heads belong to our people?' Medb said.

'Yes they do, and to the very best among them,' Ailill said.

One of their men read out the ogam on the side of the fork: that it was a single man who had thrown the fork, using one hand, and that they mustn't go past until one of them — not Fergus — did the same, single-handed.

'I am surprised,' Ailill said, 'how swiftly these four were killed.'

'That isn't what should surprise you,' Fergus said, 'but that the fork was struck from its trunk by a single stroke; that though its base is only a single cut this makes it better; and that it is driven in the way it is — for no hole was dug to receive it, and it was thrown one-handed from the back of a chariot.'

'Get rid of the obstruction for us, Fergus,' Medb said.

'Give me a chariot, then,' Fergus said. 'I'll take it out and make sure the base was made with only one cut.'

Fourteen of their chariots broke up under Fergus.

Finally he brought the fork on to dry land with his own chariot and they could see that its base was a single cut.

'We should turn our minds,' Ailill said, 'to the sort of people we are approaching. Let each of you get your food ready — it wasn't easy for you last night with the snow — and then let us hear some of the doings and stories of the sort of people we are approaching.'

It was here that they heard for the first time about the exploits of Cúchulainn.

Ailill said:

'Was it Conchobor who did this?'

'No,' Fergus said. 'He never comes to the border country without a full battle-force around him.'

'Was it Celtchar mac Uthidir, then?'

'No. He never comes to the border country, either, without a full battle-force around him.'

'Well, was it Eogan mac Durthacht?'

'No', Fergus said. 'He would never cross the border without a troop of three thousand bristling chariots around him. The man who did this deed,' Fergus said, 'is Cúchulainn. It is he who struck the branch from its base with a single stroke, and killed the four as swiftly as they were killed, and who came to the border with only his charioteer.'

'What sort of man,' Ailill said, 'is this Hound of Ulster we hear tell of? How old is this remarkable person?'

'It is soon told,' Fergus said. 'In his fifth year he went in quest of arms to the boy-troop in Emain Macha. In his seventh year he went to study the arts and crafts of war with Scáthach, and courted Emer. In his eighth year he took up arms. At present he is in his seventeenth year.'

'Is he the toughest they have in Ulster?' Medb said.

'Yes, the toughest of all,' Fergus said. 'You'll find no harder warrior against you — no point more sharp, more

swift, more slashing; no raven more flesh-ravenous, no
hand more deft, no fighter more fierce, no one of his own
age one third as good, no lion more ferocious; no barrier
in battle, no hard hammer, no gate of battle, no soldiers'
doom, no hinderer of hosts, more fine. You will find no
one there to measure against him — for youth or vigour;
for apparel, horror or eloquence; for splendour, fame or
form; for voice or strength or sternness; for cleverness,
courage or blows in battle; for fire or fury, victory, doom
or turmoil; for stalking, scheming or slaughter in the
hunt; for swiftness, alertness or wildness; and no one with
the battle-feat 'nine men on each point'— none like
Cúchulainn.'

'Let us not make too much of it,' Medb said. 'He has
only one body. He can suffer wounding. He is not beyond
being taken. Besides he is only in his early youth, and his
manly deeds are yet to come.'

'By no means,' Fergus said. 'It would be nothing strange
for him to do mighty deeds at this point. When he was
younger his acts were already manly.'

IV   CUCHULAINN'S BOYHOOD DEEDS

later plurris

HE was reared,' Fergus said, 'by his father and
mother in their oaken house on Murtheimne
Plain. There he heard great rumours about the boys in
Emain. Three times fifty boys,' Fergus said 'are always
playing in Emain. Conchobor spends one third of his
royal day watching the boys, one third playing *fidchell*,
and a third drinking ale until he falls asleep. There is no
greater warrior in Ireland,' Fergus said. 'I say it though
he drove me into exile.'

'Cúchulainn begged his mother to let him join the
boy-troop.

"You can't go," his mother said, "until there are some Ulster warriors to go with you."

"That is too long to wait," Cúchulainn said. "Point me out the way to Emain."

"Northward there," his mother said. "But it is a hard road. Sliab Fuait blocks the way."

"Still," Cúchulainn said, "I will try it."

'So he set off, with a toy shield made out of sticks and a toy javelin and his hurling-stick and ball. He kept tossing his javelin ahead and catching it again before its tail hit the ground.

'Then he ran up to Conchobor's boys without getting them to pledge his safety. He didn't know that no one went out to them on their field of play without getting a promise of safety from them.

"It is plain this young fellow is from Ulster," said Follamain, Conchobor's son, "and yet he dares us."

'They shouted at him, but still he came on against them. They flung three times fifty javelins at him, and he stopped them all on his shield of sticks. Then they drove all their hurling-balls at him, and he stopped every ball on his breast. They threw their hurling-sticks at him, three times fifty of them: he dodged so well that none of them touched him, except for a handful that he plucked down as they shot past.

'The Warp-Spasm overtook him: it seemed each hair was hammered into his head, so sharply they shot upright. You would swear a fire-speck tipped each hair. He squeezed one eye narrower than the eye of a needle; he opened the other wider than the mouth of a goblet. He bared his jaws to the ear; he peeled back his lips to the eye-teeth till his gullet showed. The hero-halo rose up from the crown of his head.

'Then he made onslaught on the boys. He laid low fifty

of them before they got to the gate of Emain. Nine of
them', Fergus said, 'flew past Conchobor and myself —
we were playing *fidchell* — and he came leaping after the
nine of them across the *fidchell* board. Conchobor caught
him by the wrist.

"These boys are being roughly handled," Conchobor
said.

"I am in the right, friend Conchobor," he said. "I left
my home, and my mother and father, to join their games,
and they treated me roughly."

"Whose son are you?" Conchobor said. "What is your
name?"

"I am Sétanta, son of Sualdam and your sister Deichtine.
I didn't expect to be hurt here."

"Well, why didn't you put yourself under the boys'
protection?" Conchobor said.

"I knew nothing about that," Cúchulainn said, "but I
ask your protection against them now."

"You have it," Conchobor said.

'Then he turned away to chase through the house
after the boy-troop.

"What are you going to do to them now?" Conchobor
said.

"Offer them my protection," Cúchulainn said.

"Promise it here and now," Conchobor said.

"I promise," Cúchulainn said.

'Then everybody went out to the play-field and the boys
who had been struck down began to get up, with the
help of their foster-mothers and fathers.'

'There was a time,' Fergus said, 'when he was a lad,
that he couldn't get to sleep at all in Emain Macha.

"Tell me, Cúchulainn, why you can't sleep here in

Emain," Conchobor said to him.

"I can't sleep unless I have the same level under my head and feet."

'So Conchobor had a block of stone brought for his head and another for his feet, and fixed a special bed between them for him.

'A while afterwards, some man went in to wake him. Cúchulainn struck him on the forehead with his fist and drove the dome of the forehead back into the brain. He knocked the stone block flat with his arm.'

'You can tell,' Ailill said, 'it was a warrior's fist, the arm of a prodigy.'

'Since that time,' Fergus said, 'no one dares to waken him, but leaves him to wake up by himself.'

'Another time he was playing ball in the playing-field east of Emain. He stood alone against the three times fifty boys. He could always beat them in every game of this kind. Once they laid hold of him, but he worked his fist on them and knocked fifty of them senseless. Then he took flight and hid under the cushion of Conchobor's bed. The whole of Ulster gathered against him — even I rose against him,' Fergus said, 'and Conchobor himself. He straightened under the bed and heaved it, bed and thirty clinging warriors, onto the floor of the house. There and then, in the house, he was encircled by Ulstermen. So,' Fergus said, 'we settled matters, and made a peace between the boy-troop and him.'

'One time Eogan Mac Durthacht challenged Ulster to battle. The men of Ulster entered the fray; Cúchulainn was left to his sleep. Ulster was beaten. Conchobor and his son Cúscraid Menn Macha, the Stammerer, were left for dead with others in heaps about them. Their wailing woke him. As he woke he stretched, and cracked the two blocks of stone near him. Bricriu, there, saw him do that,' Fergus said. 'He got up then and went to the gate of the enclosure. I met him there, in my wounds.

"Alas! God help you, friend Fergus," he said. "Where is Conchobor?"

"I don't know," I said.

'So he went out. The night was black. He made for the field of slaughter. He came upon a half-headed man who had half a corpse on his back.

"Help me, Cúchulainn," he said. "I am stricken and bear half my brother's body on my back. Carry it a while for me."

"I will not," Cúchulainn said.

'The other threw his burden at him. But he tossed it from him. They reached out at each other. And Cúchulainn was thrown down.

'Then I heard something: the Badb calling from among the corpses: "It's a poor sort of warrior that lies down at the feet of a ghost!" Cúchulainn reached up and knocked off the half-head with his hurling-stick and drove it before him, playing ball across the plain of battle.

"Is my friend Conchobor on this battlefield?"

'Conchobor made answer. Cúchulainn went toward the cry and found him in a trench, with earth piled up on all sides hiding him.

"What brings you here to the field of slaughter?" Conchobor said. "To learn what mortal terror is?"

He pulled him up from the ditch. No six of the strongest Ulstermen among us could pull so hard.

"Go before me to that house there," Conchobor said, "and light me a fire."

'He kindled a great fire for him.

"Good, so far," Conchobor said. "Now if I got a cooked pig I might come back to life."

"I'll go and get one," Cúchulainn said.

'He went out. He came upon a man by a cooking-pit in the middle of the wood, who held his weapons in one hand and cooked a boar with the other. He was a man of terrible ferocity. But Cúchulainn attacked him and took his head as well as the pig.

'Conchobor swallowed the boar.

"Back now to our house," Conchobor said.

'They found Cúscraid, Conchobor's son, on the way. He lay there heavy with wounds. Cúchulainn lifted him on his back. Then all three made for Emain Macha.'

'Another time, the men of Ulster were in their pangs. This affliction,' Fergus said 'never came to our women or our youths, or anyone not from Ulster — and therefore not to Cúchulainn or his father. None dared shed the blood of Ulstermen in this state. If they did, the pangs themselves would fall on them, or else decay, or a short life.

'Twenty-seven marauders came from the islands of Faichi. They broke into the rear enclosure as we lay in our pangs. The women there started screaming. The boy-troop heard their screams from the field of play and ran toward them. But when they saw those dark men the boy-troop took to flight, all but Cúchulainn. He attacked them with throwing-stones and his hurling-stick and killed nine of them, though they left him with fifty wounds. Then the remainder made off. What wonder that the man who did these deeds before he was five years old should cut off the heads of those four?'

'Indeed we know the boy,' Conall Cernach said. 'And I not the least; I fostered him. It wasn't long after what Fergus has told that he did another deed.

'It happened that Culann the Smith was getting ready to entertain Conchobor. He asked him not to bring too great a company, for he had no land or property to provide the feast, only what he earned by his tongs and his two hands. So Conchobor set out with only fifty chariot-fulls of the highest and mightiest of his champions to accompany him.

'First he visited the playing-field. It was his habit always, going and coming, to greet the boys and have their blessing. So it was that he saw Cúchulainn playing ball against three times fifty boys and beating them.

'When they played Shoot-the-Goal it was Cúchulainn who filled the hole with his shots and they were helpless against him. When it was their turn to shoot at the hole, all together, he turned them aside single-handed and not one ball got in. When it was time to wrestle he overthrew by himself the whole three fifties of them : and there wasn't room around him for the number needed to throw him. When they played the Stripping-Game he stripped them all stark naked. They couldn't even pluck the brooch from his cloak.

'Conchobor was amazed at this. He asked would there be the same difference in their deeds when they came to manhood. They all said there would. Conchobor said to Cúchulainn :

"Come with me," he said. "You will be a guest at this feast we are going to."

"I haven't had my fill of play yet, friend Conchobor," the boy said. "I'll follow you."

'Later when they had all arrived at the feast, Culann said to Conchobor :

"Is there anybody still to come after you?"

"No," Conchobor said, forgetting the arrangement that his foster-son was to follow them.

"I have a savage hound," Culann said. "Three chains are needed to hold him, with three men on each chain. Let him loose," he ordered, "to guard our cattle and other stock. Shut the gate of the enclosure."

'Soon the boy arrived and the hound started out for him. But he still attended to his game: he tossed his ball up and threw his hurling-stick after it and struck it; the length of his stroke never varied. Then he would cast his javelin after both, and catch it before it fell. His game never faltered though the hound was tearing toward him. Conchobor and his people were in such anguish at this that they couldn't stir. They were sure they couldn't reach him alive, even if the enclosure gate were open. The hound sprang. Cúchulainn tossed the ball aside and the stick with it and tackled the hound with his two hands: he clutched the hound's throat-apple in one hand and grasped its back with the other. He smashed it against the nearest pillar and its limbs leaped from their sockets. (According to another version he threw his ball into its mouth and so tore its entrails out.)

'Then the Ulstermen rose up to meet him, some of them over the rampart, others through the gate of the enclosure. They carried him to Conchobor's bosom. They gave a great cry of triumph, that the son of the king's sister had escaped death.

'Culann stood in his house.

"You are welcome, boy, for your mother's heart's sake. But for my own part I did badly to give this feast. My life is a waste, and my household like a desert, with the loss of my hound! He guarded my life and my honour," he said; "a valued servant, my hound, taken from me. He was shield and shelter for our goods and herds. He guarded all our beasts, at home or out in the fields."

"That doesn't matter," the boy said. "I'll rear you a pup from the same pack. Until that hound grows up to do his work, I will be your hound, and guard yourself and your beasts. And I will guard all Murtheimne Plain. No herd or flock will leave my care unknown to me."

"Cúchulainn shall be your name, the Hound of Culann," Cathbad said.

"I like that for a name!" Cúchulainn said.

'What wonder that the man who did this at the end of his sixth year should do a great deed at the present time when he is full seventeen?' Conall Cernach said.

'There was another deed he did,' Fiacha Mac Fir Febe said. 'Cathbad the druid was staying with his son, Conchobor mac Nesa. He had one hundred studious men learning druid lore from him — this was always the number that Cathbad taught.

'One day a pupil asked him what that day would be lucky for. Cathbad said if a warrior took up arms for the first time that day his name would endure in Ireland as a word signifying mighty acts, and stories about him would last forever.

'Cúchulainn overheard this. He went to Conchobor and claimed his weapons. Conchobor said:

"By whose instruction?"

"My friend Cathbad's," Cúchulainn said.

"We have heard of him," Conchobor said, and gave him shield and spear. Cúchulainn brandished them in the middle of the house, and not one piece survived of the fifteen sets that Conchobor kept in store for new warriors or in case of breakage. He was given Conchobor's own weapons at last, and these survived. He made a flourish and saluted their owner the king and said:

"Long life to their seed and breed, who have for their king the man who owns these weapons."

'It was then that Cathbad came in and said:

"Do I see a young boy newly armed?"

"Yes," Conchobor said.

"Then woe to his mother's son," he said.

"What is this? Wasn't it by your own direction he came?" Conchobor said.

"Certainly not," Cathbad said.

"Little demon, why did you lie to me?" Conchobor said to Cúchulainn.

"It was no lie, king of warriors," Cúchulainn said. "I happened to hear him instructing his pupils this morning south of Emain, and I came to you then."

"Well," Cathbad said, "the day has this merit: he who arms for the first time today will achieve fame and greatness. But his life is short."

"That is a fair bargain," Cúchulainn said. "If I achieve fame I am content, though I had only one day on earth."

'Another day came and another druid asked what that day would be lucky for.

"Whoever mounts his first chariot today," Cathbad said, "his name will live forever in Ireland."

Cúchulainn overheard this also, and went to Conchobor and said:

"Friend Conchobor, my chariot!"

'A chariot was given to him. He clapped his hand to the chariot between the shafts, and the frame broke at his touch. In the same way he broke twelve chariots. At last they gave him Conchobor's chariot and that survived him.

'He mounted the chariot beside Conchobor's charioteer. This charioteer, Ibor by name, turned the chariot round where it stood.

"You can get out of the chariot now," the charioteer said.

"You think your horses are precious," Cúchulainn said, "but so am I, my friend. Drive round Emain now, and you won't lose by it."

'The charioteer set off.

'Cúchulainn urged him to take the road to the boy-troop, to greet them and get their blessing in return. After this he asked him to go further along the road. Cúchulainn said to the charioteer as they drove onward:

"Use your goad on the horses now."

"Which direction?" the charioteer said.

"As far as the road will take us!" Cúchulainn said.

'They came to Sliab Fuait. They met Conall Cernach there — for to Conall Cernach had fallen the care of the province boundary that day. Each of Ulster's heroic warriors had his day on Sliab Fuait, to take care of every man who came that way with poetry, and to fight any others. In this way everyone was challenged and no one slipped past to Emain unnoticed.

"May you prosper," Conall said. "I wish you victory and triumph."

"Conall, go back to the fort," Cúchulainn said, "and let me keep watch here a little."

"You would do for looking after men of poetry," Conall said. "But you are a little young still for dealing with men of war."

"It might never happen at all," Cúchulainn said. "Let us wander off, meanwhile," he said, "to view the shore of Loch Echtra. Warriors are often camped there."

"It is a pleasant thought," Conall said.

'They set off. Suddenly Cúchulainn let fly a stone from his sling and smashed the shaft of Conall Cernach's chariot.

"Why did you cast that stone, boy?" Conall said.

"To test my hand and the straightness of my aim," Cúchulainn said. "Now, since it is your Ulster custom not to continue a dangerous journey, go back to Emain, friend Conall, and leave me here on guard."

"If I must," Conall said.

'Conall Cernach wouldn't go beyond that point.

'Cúchulainn went on to Loch Echtra but found no one there. The charioteer said to Cúchulainn that they ought to go back to Emain, that they might get there for the drinking.

"No," Cúchulainn said. "What is that peak there?"

"Sliab Mondairn," the charioteer said.

"Take me there," Cúchulainn said.

'They travelled on until they got there. On arriving at the mountain, Cúchulainn asked:

"That white heap of stones on the mountain-top, what is it called?"

"The look-out place, Finncarn, the white cairn," the charioteer said.

"That plain there before us?" Cúchulainn said.

"Mag mBreg, Breg Plain," the charioteer said.

'In this way he gave the name of every fort of any size between Temair and Cenannos. And he recited to him also all fields and fords, all habitations and places of note, and every fastness and fortress. He pointed out at last the fort of the three sons of Nechta Scéne, who were called Foill (for deceitfulness) and Fannall (the Swallow) and Tuachell (the Cunning). They came from the mouth of the river Scéne. Fer Ulli, Lugaid's son, was their father and Nechta Scéne their mother. Ulstermen had killed their father and this is why they were at enmity with them.

"Is it these who say," Cúchulainn said, "that they have killed as many Ulstermen as are now living?"

"They are the ones," the charioteer said.

"Take me to meet them," Cúchulainn said.

"That is looking for danger," the charioteer said.

"We're not going there to avoid it," Cúchulainn said.

'They travelled on, and turned their horses loose where bog and river met, to the south and upstream of their enemies' stronghold. He took the spancel-hoop of challenge from the pillar-stone at the ford and threw it as far as he could out into the river and let the current take it — thus challenging the ban of the sons of Nechta Scéne.

'They took note of this and started out to find him.

'Cúchulainn, after sending the spancel-hoop downstream, lay down by the pillar-stone to rest, and said to his charioteer:

"If only one man comes, or two, don't wake me, but wake me if they all come."

'The charioteer waited meanwhile in terror. He yoked the chariot and pulled off the skins and coverings that were over Cúchulainn, trying not to wake him, since Cúchulainn had told him not to wake him for only one.

'Then the sons of Nechta Scéne came up.

"Who is that there?" said one.

"A little boy out in his chariot today for the first time," the charioteer said.

"Then his luck has deserted him," the warrior said. "This is a bad beginning in arms for him. Get out of our land. Graze your horses here no more."

"I have the reins in my hand," the charioteer said.

Then Ibor said to the warrior:

"Why should you earn enmity? Look, the boy is asleep."

"A boy with a difference!" cried Cúchulainn. "A boy who came here to look for fight!"

"It will be a pleasure," the warrior said.

"You may have that pleasure now, in the ford there," Cúchulainn said.

"You would be wise," the charioteer said, "to be careful of the man who is coming against you. Foill is his name," he said. "If you don't get him with your first thrust, you may thrust away all day."

"I swear the oath of my people that he won't play that trick on an Ulsterman again when my friend Conchobor's broad spear leaves my hand to find him. He'll feel it like the hand of an outlaw!"

'He flung the spear at him, and it pierced him and broke his back. He removed the trophies, and the head with them.

"Watch this other one," the charioteer said. "Fannall is his name, and he treads the water no heavier than swan or swallow."

"I swear he won't use that trick on an Ulsterman again," Cúchulainn said. "You have seen how I foot the pool in Emain."

'They met in the ford, and he killed the man and took away the trophies and the head.

"Watch this next one advancing against you," the charioteer said. "Tuachell is his name, and he wasn't named in vain. He has never fallen to any weapon."

"I have the *del chliss* for him, a wily weapon to churn him up and red-riddle him," Cúchulainn said.

'He threw the spear at him and tore him asunder where he stood. He went up and cut off his head. He gave the head and trophies to his charioteer.

'Then a scream rose up behind them from the mother, Nechta Scéne. Cúchulainn lifted the trophies off the ground and brought the three heads with him into the chariot, saying:

"I won't let go of these trophies until we reach Emain Macha."

"They set out for Emain Macha with all his spoils. Cúchulainn said to his charioteer:

"You promised us great driving. We'll need it now after our fight, with this chase after us."

'They travelled onward to Sliab Fuait. So fleet their haste across Breg Plain, as he hurried the charioteer, that the chariot-horses overtook the wind and the birds in flight, and Cúchulainn could catch the shot from his sling before it hit the earth.

'When they got to Sliab Fuait they found a herd of deer before them.

"What are those nimble beasts there?" Cúchulainn said.

"Wild deer," the charioteer said.

Cúchulainn said:

"Which would the men of Ulster like brought in, a dead one or a live one?"

"A live one would startle them more," the charioteer said. "It isn't everyone who could do it. Every man there has brought home a dead one. You can't catch them alive."

"I can," Cúchulainn said. "Use your goad on the horses, over the marsh."

'The charioteer did so until the horses bogged down. Cúchulainn got out and caught the deer nearest to him, the handsomest of all. He lashed the horses free of the bog and calmed the deer quickly. Then he tethered it between the rear shafts of the chariot.

'The next thing they saw before them was a flock of swans.

"Would the men of Ulster prefer to have these brought

in alive or dead?" Cúchulainn said.

"The quickest and the most expert take them alive," the charioteer said.

'Cúchulainn immediately flung a little stone at the birds and brought down eight of them. Then he flung a bigger stone that brought down twelve more. He did this with his feat of the stunning-shot.

"Gather in our birds now," Cúchulainn said to his charioteer. "If I go out to them this wild stag will turn on you."

"But it's no easier if I go," the charioteer said. "The horses are so maddened that I can't get past them. I can't get over the two iron rims of the chariot wheels, they are so sharp. And I can't get past the stag; his antlers fill all the space between the chariot's shafts."

"Step out onto the antlers," Cúchulainn said. "I swear the oath of Ulster's people, I'll turn my head on him with such a stare, I'll fix him with such an eye, that he won't dare to stir or budge his head at you."

'He did this. Cúchulainn tied the reins and the charioteer gathered up the birds. Then Cúchulainn fastened the birds to the cords and thongs of the chariot. It was in this manner that they came back to Emain Macha: a wild stag behind the chariot, a swan-flock fluttering above, and the three heads of Nechta Scéne's sons inside the chariot.

'They came to Emain.

"A man in a chariot advancing upon us," cried the watcher in Emain Macha. "He'll spill the blood of the whole court unless you see to him and send naked women to meet him."

'Cúchulainn turned the left chariot-board toward Emain in insult, and he said:

"I swear by the oath of Ulster's people that if a man

isn't found to fight me, I'll spill the blood of everyone in this court.''

"Naked women to him!" Conchobor said.

'The women of Emain went forth, with Mugain the wife of Conchobor mac Nesa at their head, and they stripped their breasts at him.

"These are the warriors you must struggle with today," Mugain said.

'He hid his countenance. Immediately the warriors of Emain seized him and plunged him in a vat of cold water. The vat burst asunder about him. Then he was thrust in another vat and it boiled with bubbles the size of fists. He was placed at last in a third vat and warmed it till its heat and cold were equal. Then he got out and Mugain the queen gave him a blue cloak to go round him with a silver brooch in it, and a hooded tunic. And he sat on Conchobor's knee, and that was his seat ever after.

'What wonder,' Fiacha mac Fir Febe said, 'that the one who did this in his seventh year should triumph against odds and beat his match today, when he is fully seventeen years old?'

V   'DEATH   DEATH!'

'WELL' Ailill said, 'let us be off.' They went to Mag Muceda, the Pig-keeper's Plain, and there Cúchulainn cut down an oak tree in their path and cut an ogam message into its side. He wrote there that no one was to pass that oak until a warrior had leaped it in his chariot at the first attempt. So they pitched their tents and set themselves to leaping the tree in their chariots. Thirty horses fell on that spot, and thirty chariots were smashed there, and the place has

been called Belach nAne ever since, the Pass where they
Drove.

They waited there until morning. Then Fraech mac Fidaig
was called and Medb said to him :
   'Fraech, we need your help to clear this nuisance away.
Go and find Cúchulainn and challenge him.'
   Early in the morning he went out with nine others, and
they came to Ath Fuait. They saw the boy there, washing
in the river.
   'Wait here,' Fraech said to his followers. 'I'll attack
him there in the water; he isn't good in water.'
   He stripped off his clothes and went up to him in the
water.
   'If you came any nearer,' Cúchulainn said, 'I would
have to kill you, and that would be a pity.'
   'All the same, I'm coming to meet you in the water,'
Fraech said. 'You'll have to fight.'
   'Choose your style of combat then,' Cúchulainn said.
   'Each to keep one arm round the other,' Fraech said.
   They grappled a long time in the water until Fraech
went under. Cúchulainn pulled him up again.
   'Now,' Cúchulainn said, 'will you let me spare you?'

'I wouldn't have that said,' Fraech said.

Cúchulainn thrust him down once more and Fraech perished. Ath Froich is the name of that ford still. After he had been laid on the dry land, his followers carried his body to the camp and the whole company lamented the warrior Fraech. Then they saw a troop of women in green tunics gather about the body of Fraech mac Fidaig and bear him away into the *síd*. Síd Froich is the name of the hill since that time.

Then Fergus leaped in his chariot across the oak tree.

Some say they went from here to Ath Meislir, where Cúchulainn slew six of the host, Meslir and the others. Others say they went to Ath Taiten and that the six Cúchulainn slew were six Dúngals of Irros.

They left then for Fornocht, the Naked Place. Medb had a young hound. Its name was Baiscne. Cúchulainn slung a stone at it and took off its head. The place where this happened is called Druim Baiscne since that time, the Ridge of Baiscne.

'Shame on you all,' Medb said, 'not to be out after this pestering demon that is killing you all!'

They took off in pursuit of him then, until their chariot shafts broke. Next day they crossed a high place, Cúchulainn roaming ahead of them.

At the place called Támlachtai Orláim, Orlám's Burial Mark, a little to the north of the sanctuary Dísert Lochait, Cúchulainn came upon a charioteer cutting wood-shafts. He was Orlám's charioteer; Orlám was a son of Ailill and Medb. Some say that the shaft of Cúchulainn's own chariot had broken, and that he also was cutting out a new one when he saw Orlám's charioteer.

'This is madness,' Cúchulainn said, thinking it was an

Ulster warrior. 'Are there Ulstermen here, with an attacking army coming up behind them?'

He went up to stop the charioteer. He watched him for
a while cutting out wood for a chariot-shaft.

'What are you doing here?' Cúchulainn said.

'Getting chariot-shafts,' the charioteer said. 'We smashed our chariots chasing that wild deer Cúchulainn. Help
me with them,' he said. 'Would you rather cut out the
shafts or do the trimming?'

'I'll do the trimming,' Cúchulainn said.

Then, under the other's eyes, he stripped the holly-
shafts through his clutched fist, paring them clean, knot
and bark. The charioteer said in fright:

'This isn't your usual work.'

'Who are you?' Cúchulainn said.

'Orlám's charioteer. He is a son of Ailill and Medb. And
who are you?' the charioteer said.

'Cúchulainn,' he said.

'Alas!' the charioteer said.

'You needn't worry,' Cúchulainn said. 'Where is your
master?'

'He is over there by the dike,' the charioteer said.

'Come with me, now,' Cúchulainn said. 'I have no
quarrel with charioteers.'

Cúchulainn went up to Orlám and slew and beheaded
him, and shook the head at the host. He set the head on
the charioteer's back and said:

'Take this with you and keep it like that all the way
into the camp. If you do anything but exactly what I say
you'll get a shot from my sling.'

The charioteer went up close to the camp and took the
head from his back, and told Medb and Ailill his story.

'This isn't like catching birds,' she said.

'And he told me,' the charioteer said, 'that if I didn't

take it on my back all the way into the camp he'd break my head with a stone.'

Orlám's charioteer was standing at this time between Ailill and Medb outside the camp. Cúchulainn hurled a stone at him, shattering his head so that the brains spattered the ears. His name was Fertedil. It is not true, therefore, that Cúchulainn didn't kill charioteers; he killed them if they did wrong.

The three sons of Gárach were waiting at the ford which now bears their name, Ath meic Gárach. These are their names: Lon, Ualu and Diliu — the blackbird, the prideful, the torrent. The three charioteers were there also, the three foster-sons, Meslir, Meslaech and Meslethan. They couldn't bear the thought of Cúchulainn killing two foster-sons of the king, and a son by blood, and shaking the head at the host. They planned to kill Cúchulainn themselves and lift the scourge from the army. Three shafts of aspen were cut for their charioteers so that all six could go against him together, thus breaking the rule of fair fight. But he slew them all.

Cúchulainn swore an oath in Methe that from this time on, whenever he laid eyes on Ailill and Medb, he would hurl a sling-stone at them. It was then he shot a sling-stone south across the ford and killed Medb's squirrel as it sat close to her neck. Hence comes Méthe Tog, Squirrel Neck, as the name of that place. He killed also a pet bird perching close to Ailill's neck; from which comes Ath Méthe nEuin, or Bird Neck Ford. Others say that the bird and the little squirrel were both perched on Medb's shoulders when their heads were torn off by the sling-stones. At this time also Reuin was drowned in the lake now called after him.

'He can't be far off,' Ailill said one time to his sons, the
Maine. They rose up, looking about them. As they were
settling down again, Cúchulainn struck one of them,
shattering his head.

'That was a fine way to rise against him,' Maenén the
jester said, 'after all your boasting! I would have knocked
his head off.'

At which a stone from Cúchulainn shattered his head
also. The following, then, is the list of the slain: Orlám,
firstly, on the hill that bears his name; Fertedil between
two protectors; the three sons of Gárach on their ford;
and Maenén on his hill.

'I swear by the god of my people,' Ailill said, 'I'll cut
in two any man who scoffs at Cúchulainn from now on.
Let us be off now, travelling day and night,' he said, 'until
we get to Cuailnge. The man will kill two thirds of our
army if he goes on like this.'

Then the magical sweet-mouth harpers of Caín Bile came
out from the red cataract at Es Ruaid, to charm the
host. But the people thought that these were spies from
Ulster coming among them, and they gave chase after
them until they ran in the shape of deer far ahead of
them to the north among the stones at Liac Mór, they
being druids of great knowledge.

Lethan stood at the ford on the river Níth in Conaille, in
a rage at what Cúchulainn had done, and waited for him.
But Cúchulainn cut off his head and left it with the body.
The ford on the Níth is named Ath Lethan from this. In
the previous ford so many chariots were shattered in the
fighting that it is still called Ath Carpat, Ford of Chariots.
On the shoulder of land that lies between these fords,
Mulca, Lethan's charioteer, fell. Hence its name is Guala

Mulchai, Mulcha's Shoulder. In this manner, as the armies crossed Breg Plain, he sent men continually to their graves.

Now it was that the Morrígan settled in bird shape on a standing stone in Temair Chuailnge, and said to the Brown Bull :

> 'Dark one are you restless
> > do you guess they gather
> to certain slaughter
> > the wise raven
> groans aloud
> > that enemies infest
> the fair fields
> > ravaging in packs
> ıearn I discern
> > rich plains
> softly wavelike
> > baring their necks
> greenness of grass
> > beauty of blossoms
> on the plains war
> > grinding heroic
> hosts to dust
> > cattle groans the Badb
> the raven ravenous
> > among corpses of men
> affliction and outcry
> > and war everlasting
> raging over Cuailnge
> > death of sons
> death of kinsmen
> > death   death !'

Thereupon the Bull moved to Sliab Cuilinn with his fifty heifers and his herdsman Forgaimen driving him. He threw off the three times fifty boys who always played on his back, killing two thirds of them, and he tore up a trench through the land of Marcéni in Cuailnge, tossing the earth back over him with his heels.

From the gloomy waters of Saili Imdorchi, in the district of Conaille, until they reached Cuailnge, Cúchulainn killed no one, being then at Sliab Cuinciu. There he swore again that whenever he caught sight of Medb he would hurl a sling-stone at her head. This was no easy thing to do, for Medb never went about unless she was protected by half her army holding a barrel-shaped shelter of shields over their heads. One of Medb's bondmaids named Lochu went to fetch water, surrounded by a troop of many women, and Cúchulainn mistook her for Medb. He shot two stones at her from Cuinciu and killed her on the plain in the place known as Réid Locha, Lochu's level ground, in Cuailnge.

VI   FROM FINNABAIR CHUAILNGE TO CONAILLE

IT IS SAID in one version of the tale that the armies divided at Finnabair in Cuailnge and laid waste the country with fire. They rounded up all the women and boys and girls and cattle in Cuailnge and brought them to Finnabair.

'You haven't done well enough yet,' Medb said. 'I don't see the bull with you.'

'There is no trace of him anywhere in the province,' they said.

One of Medb's herdsmen, Lóthar, was summoned.

'Where do you think the bull might be?' she said.

'I tremble to tell you,' the herdsman said, 'but on the

same night that the men of Ulster were laid low by their pangs, he left with all his three score heifers. He is now in Dubchoire, the Black Cauldron, in Glenn Gat of the Osiers.'

'Make yourselves ready,' Medb said. 'Take a shackle of osiers between each pair of you, and catch him.'

They did as she said, and hence the name of this place is Glenn Gat, the Valley of Osiers. They encircled the bull there and drove him toward Finnabair. And there he saw the cowherd Lóthar and attacked him in a fury, taking out his entrails on the horns. He attacked the camp with his three score heifers, and fifty heroes perished in his path. Then the bull vanished out of the camp and, to their shame, no one could say where he had gone. Medb asked the herdsmen where the bull might be.

'He is back in the fastnesses of Sliab Cuilinn.'

They headed for that place, ravaging Cuailnge as they went, but they couldn't find the bull there.

It is further said, in this version, that the river Cronn rose up against them to the height of the treetops and they had to pass the night by the edge of the water. In the morning Medb ordered some of her followers across it. The famous warrior Ualu tried it. To cross the river he shouldered a big flagstone so that the water wouldn't force him backward. But the river overwhelmed him, stone and all, and he drowned. His grave, with his stone, is on the roadway by the river; Lia Ualonn is its name, Ualu's Flagstone. It is there that Cúchulainn killed Cronn and Caemdele in heroic fury; and a further hundred warriors died as they struggled, together with Roan and Roae, the two chroniclers of the Táin. Some say that this is the reason the tale of the Táin was lost and had to be found again long afterward. One hundred and twenty-four kings died by his hand at the same river.

So they went upward along the river Cronn until they reached its source. They were crossing between the spring and the mountain-summit when Medb called them back. She chose to cross the summit itself and mark their track forever as a mark of dishonour to Ulster. It took them three days and three nights, tearing up the earth before them, to form the gap Bernas Bó Cuailnge.

Then they went over Bernas Bó Cuailnge with all their cattle and belongings and they passed the night in Glenn Dáilimda in Cuailnge, at the place now called Botha after the huts they made there. Next day they travelled to the river Colptha. Recklessly they tried a crossing, but it too rose against them and bore off a hundred of their charioteers toward the sea. Cluain Carpat, the Chariot-Meadow, is the name of the place where they drowned. They had to move along the river Colptha up to its source, then to Bélat Aliuin. They passed the night between Cuailnge and Conaille at Liasa Liac, so named because the armies built stone shelters for their calves there. After this they went across Glenn Gatlaig, but the river Gatlaig rose up against them also. Previously it was called Sechaire, but it is known as the river Gatlaig since that time, after the osiers they carried their calves in. They spent the night in Druim Féne in Conaille. Such, then, according to one version were their travels from Cuailnge to Conaille Plain.

However, other writers of this tale, and other books, treat events differently from the dividing up of the armies at Finnabair to the arrival in Conaille. Thus, when all had brought their spoils back with them to Finnabair in Cuailnge, Medb said :

'Divide up the armies. Our forces can't all advance on

the one road. Ailill can take half of them by the Midluachair road. We'll go with Fergus by way of Bernas Bó
Ulad.'

'That leaves us the difficult half of the army,' Fergus
said. 'We'll have to cut a gap to get the cattle over the
mountain.'

That is what they did, and such is the origin, says this
author, of the name Bernas Bó Ulad.

It is at this time that Ailill took his charioteer Cuillius
aside.

'Watch Medb and Fergus today for me. I don't know
why they are so intimate and I want you to find me some
sign.'

Then Cuillius found the couple together at Cluithre,
where they had lingered behind as the army moved on.
Cuillius moved closer. They didn't hear him spying on
them. It happened that Fergus's sword was laid down
close by him. Cuillius drew it out of its sheath, leaving
the sheath empty. Then he went back to Ailill.

'Well,' Ailill said.

'Well indeed,' Cuillius said. 'Here is your sign. I dis-
covered them sleeping together as you thought.'

'Fair enough,' Ailill said, and they grinned at each
other. 'It is all right,' Ailill said. 'She is justified. She does
it to keep his help on the Táin. Now, keep the sword in
good order. Put it under your chariot-seat with a piece of
linen around it.'

Meanwhile Fergus was looking about for his sword.

'This is terrible,' he said.

'What is wrong?' Medb said.

'The wrong I have done Ailill,' Fergus said. 'Wait here.
I must go into the wood. Don't be surprised if I am gone
a while.'

Medb didn't understand that his sword had vanished.

He left her, taking his charioteer's sword with him, and cut a wooden sword from a tree. This is how Fid Mór-thruaille, the Wood of the Great Scabbard, in Ulster, got its name.

'Let us get back to our companions,' Fergus said.

All the armies met in the plain and made camp. Ailill sent for Fergus to play *fidchell* with him. When Fergus entered the tent Ailill started laughing at him.

Fergus said:
> 'Better be laughed at
> > mad after the act
> my sword top maddened
> > Macha's curse quick doom
> Galeóin swords outcry
> > women unvanquished
> dark driven to meet them
> > spear flock sword flock
> among leaders of armies
> > on Nes's boy's hill
> armies struggle in fury
> > men's severed necks.'

Ailill said:
> 'Why so wild
> > without your weapon
> on heights of a certain
> > royal belly
> in a certain ford
> > was your will worked
> or your heroism
> > an empty shout
> to Medb's oaths
> > tribes of men

can bear witness
      sucked dry in the struggle
with giddy women
      crawling entering
battling with great
      murky deeds
under cover
      everywhere.
'Now sit down,' Ailill said, 'and we will play *fidchell*.
You are very welcome.
    'You play *fidchell* and *buanbach*
      with a king and queen
ruling the game
      their eager armies
in iron companies
      all around them
not even if you win
      can you take my place
I know all
      about queens and women
I lay first fault
      straight at women's
own sweet swellings
      and loving lust
valorous Fergus
      coming and going
with cattle bellowings
      and huge forces
all over Finnabair's
      rich places
in kingly form
      with fire of dragon
hiss of snake
      blow of lion

thrusting out in front
            Roech's son Fergus
grandson of Rus
            the king of kings.'
They began their game of *fidchell*, advancing the gold
and silver men over the bronze board.
    Ailill said :
            'It isn't right
                    that death should take
            this sweet slight king
                    on the coppery point
            the handsomer
                    on this mad board
            mighty Medb
                    the less secure
            these wise men
                    I move against Fergus
            let right be done
                    as our game goes.'
    Medb said :
            'Hold a while
                    your clownish words
            don't forget
                    what still remains
            with the gentle boy troop
                    all might change
            a wise judge
                    bears no grudge
            have no more
                    to do with those
            who keep their cattle
                    with a vengeance
            men's eyes downcast
                    and Fergus cleared.'

Fergus said:
>   'A pity friend
>   we hack each other
>   with sharp words
>   in the public gaze
>   right speech offends
>   right ways run wrong
>   javelins wiped
>   kings killing kings
>   at a great man's word.'

They stayed there that night. Next morning Ailill
spoke:
>   'One warrior out
>   before huge armies
>   by Nes's Cronn water
>   his deeds loosed
>   at the men of Connacht
>   men's blood floods
>   from hacked necks
>   great men's deaths
>   dark driven to meet
>   waves mounting up
>   where the beardless hero
>   comes from Ulster.'
Medb said:
>   'Don't call down violence
>   mighty Mac Mata
>   chariot onslaughts
>   from rocky heights
>   men massing
>   women carried off
>   cattle before them
>   and the heads of armies

       swords smashed
            on either side
       men's deeds of battle
            wrought in the murk
       oxen driven
            women stolen
       great armies turning
            from the battle plain
       of Cuailnge
            now the army sleeps.'
Fergus said :
       'Huge heads stuck
            on chariot prow
       and gibbet face
            great hearted heroes
       will swear by their people
            squabble over queens
       forge at the foe.'
Medb said :
       'As you have said
            so let it be
       let it be so
            he bends to your yoke
       hordes are marching
            Ailill's power
       put in your hands
            to what effect.'

They moved onward as far as the Cronn river. Maine,
one of Ailill's sons, spoke to them :
       'Send me out swiftly
            mother father
       fair deeds done
            for the horned herds

stand fast till I get
    in chariot reach
away from the herds
    and the battle field
in mighty acts
    is swept clean.'
Fergus said:
    'Mighty son
        don't venture out
    it is only asking
        to have your head
    knocked from your neck
        by the boy with no beard
    who comes from the heights
        howls on the plain
    summons up rivers
        shakes the woods
    wrenches into shapes
        mighty acts
    men in great numbers
        drowned in the waters
    Ailill hurt
        and Medb mocked
    faces cast down
        in the bristling battle.
'Let me travel ahead with the exiles,' Fergus said, 'to
make sure there is no foul play against the boy. Send the
cattle in front and the armies after us, and the women in
the rear.'
    Then Medb said:
    'On your soul and oath
        Fergus listen
    guard these cattle
        with your good armies

       in conquering rage
            halt the men of Ulster
       or a roar will rise
            on the Plain of Ai
       rise overcome
            and we'll meet again
       on the army's track.'
Fergus said:
       'Spare us Medb
            your shameless talk
       and harassment
            in the public gaze
       no limp soft son
            was ever mine
       at the struggle in Emain
            I'll strike my people
       no more blows
            let me out
       from under your weight
            no man come breathing
       down my neck
            to do your work
       on another outing.'

At a ford on the Cronn river Cúchulainn came to meet
them.

    'Laeg, my friend,' he said to his charioteer, 'the army
is upon us.'

    The charioteer said:
       'I swear to the gods
            I'll do great deeds
       before these warriors
            driving to triumph

> at full force
> > on slender steeds
> with yokes of silver
> > and golden wheels
> to crush kings' heads
> > my driven steeds
> will take us leaping
> > to victory.'

Cúchulainn said:
> 'Now friend Laeg
> > set our course headlong
> into the crush
> > for Macha's great triumph
> let them stray like women
> > on the plain in terror
> the teams' heads set
> > against Ailill and Medb
> through two armies
> > like placid herds
> grinding among them
> > our vengeful path.

'I summon the waters to help me,' Cúchulainn said. 'I summon air and earth; but I summon now above all the Cronn river:

> 'Let Cronn itself fall-to in the fight
> to save Murtheimne from the enemy
> until the warrior's work is done
> on the mountain-top of Ochaine.'

And the water reared up to the treetops.

Then the Maine, son of Ailill and Medb, went out before all. Cúchulainn slew him in the ford and thirty horsemen of his company were overwhelmed with him. Later

Cúchulainn slew another thirty-two warriors in the water.

They pitched their tents at this ford. Lugaid mac Nois Allchomaig went out with thirty horsemen to talk with Cúchulainn.

'Lugaid, I bid you welcome,' Cúchulainn said. 'If a flock of wild birds were grazing on Murtheimne Plain now, I'd give you one and share another; if the salmon were swimming the weirs or river-mouths now I'd give you one and share another, with the three proper herbs: cress of the stream, marshwort and sea-herb. And I would stand for you in the ford of battle.'

'I believe you, beloved son,' Lugaid said. 'I wish you a wealth of followers.'

'You have a fine army,' Cúchulainn said.

'You could hold them single-handed,' Lugaid said.

'If it was one by one the army came against me, your Ulster enemy wouldn't disgrace you, Lugaid,' Cúchulainn said. 'I have right and might to sustain me. Friend Lugaid,' he said, 'do the hosts fear me?'

'I swear by the gods,' Lugaid said, 'they daren't make water in ones or twos outside the camp, but have to go in twenties and thirties.'

'I have something new for them,' Cúchulainn said. 'I am taking up sling-throwing. Tell me now, Lugaid, what you want.'

'That you will spare my own men,' Lugaid said.

'You have my promise, provided you point them out to me by a sign. And tell my friend Fergus to show a sign among his men too. And tell the healers to show themselves by a sign — but they must swear to watch over my life and send me food every night.'

Lugaid returned. He found Fergus in Ailill's tent and he called him out and told him the news.

Then they heard Ailill:
>   'What are you whispering
>   this is no sportfield
>   for our great army
>   he chooses among us
>   for the sake of Roech's son
>   who plays king in our place
>   as we hear tell
>   though we get great help
>   through Medb's sweet needs
>   let us take our few men
>   to the favoured tents
>   and all be safe
>   from flying flagstones
>   and hurtling sods
>   by these secret meetings
>   I know he is near.'

'I swear by the gods I can't promise that without asking the boy again,' Lugaid said.

'Lugaid,' Fergus said, 'will you do this for me? Go and ask him to let me take Ailill and his troop of three thousand among my own men. Bring an ox, a salt pig and a barrel of wine with you.'

Lugaid went and asked him.

'It is all the same to me where he goes,' Cúchulainn said.

So the two troops mingled together and they stayed so for the night — or for twenty nights, or thirty, as some say. But even so Cúchulainn destroyed thirty of Ailill's warriors with his sling.

'Things are growing worse for you,' Fergus said. 'The men of Ulster will soon rise from their pangs, and then they'll grind you to grit and gravel. Besides, this is a bad place to fight.'

He set out then toward Cúil Airthir, in the east.

Cúchulainn slew thirty warriors at Ath Duirn, the Ford of the Fist, and they couldn't reach Cúil Airthir until night came. Cúchulainn slew thirty more of them there before they pitched their tents. In the morning Ailill's charioteer Cuillius was washing the bands of his chariot-wheels in the ford and Cúchulainn struck him with a stone and killed him; from which comes the name Ath Cuillne, the Ford of Cuillius in Cúil Airthir. They pressed on then, reaching Druim Féne in Conaille for the night — and that is the second version of how they reached that place.

VII   SINGLE COMBAT

CUCHULAINN continued to harass them

there. He slew a hundred men on each of the three nights they stayed in that place, plying the sling on them from the hill Ochaine nearby.

'At this rate,' Ailill said, 'our army will melt away at his hands. Bring Cúchulainn this offer: I to give him a part of Ai Plain equal to the whole plain of Murtheimne, with the best chariot to be found at Ai, and harness to equip a dozen men. Or offer him, if he would like it more, his native plain, with twenty-one bondmaids and compensation for anything of his — cattle or household goods — that we have destroyed. He for his part to take service under me, who am more worthy of him than the half-lord he serves now.'

'Who will take this message?'

'Mac Roth there.' (Mac Roth could circle the whole of Ireland in one day.)

Mac Roth set out to Delga with the message from

Ailill and Medb; it is there in Delga that Fergus thought
Cúchulainn might be found. A heavy snow fell that night,
that turned all the provinces of Ireland into a snow-white
plain.

'There is a man coming,' Laeg said to Cúchulainn.
'He has a linen band round his yellow hair. He grasps
a wrathful club. An ivory-hilted sword hangs at his waist.
A red-embroidered hooded tunic is wrapped around him.'

'Which of the kings' warriors is he?' Cúchulainn said.

'A dark, good-looking, broad-faced man, with a bronze
brooch in his handsome brown cloak, a tough triple shirt
next his skin, and a pair of well-worn shoes between his
feet and the ground. He holds a peeled hazel-wand in one
hand and a single-edged sword with guards of ivory in the
other.'

'Those are the marks of a herald,' Cúchulainn said.

Mac Roth asked Laeg whose servant in arms he was.

'That man's there,' Laeg said.

Cúchulainn was squatting haunch-deep in the snow,
stripped and picking his shirt. Mac Roth asked him whose
servant in arms he was.

'I serve Conchobor mac Nesa,' Cúchulainn said.

'Can you say no clearer than that?'

'It's clear enough,' Cúchulainn said.

'Where can I find Cúchulainn?' Mac Roth said.

'What have you to say to him?' Cúchulainn said.

Mac Roth gave him the full message.

'If Cúchulainn were here he wouldn't sell his mother's
brother for another king.'

Mac Roth came to Cúchulainn again and said they
would give him the noblest women and all the milkless
cattle out of their plunder if he would stop using his sling
against them at night — he might kill as he chose by day.

'I can't agree to that,' Cúchulainn said, 'for if you take

away the bondwomen our freewomen will have to take
to the grinding-stones, and if you take away our milch
cows we would have to go without milk.'

Mac Roth came to Cúchulainn again, and said they
would leave him instead the bondwomen and the milch
cows.

'I can't agree to that either,' Cúchulainn said, 'for the
men of Ulster would sleep with the bondwomen and
beget slavish sons, and they would use the milch cows
for meat in the winter.'

'Is there anything that will do?' the messenger said.

'There is,' Cúchulainn said, 'but I won't say what it is.
If you can find anyone who knows what I mean, I'll agree
to it.'

'I know what he has in mind,' Fergus said, 'and indeed
it bodes you no good. This is his plan : that he will fight
you one by one in the ford, and that no cattle will be
taken from the ford for a day and a night after each
combat. This plan will gain time for him until help
comes from the men of Ulster — and I am amazed,'
Fergus said, 'that they are so long recovering from their
pangs.'

'It will be easier on us, no doubt,' Ailill said, 'to lose
one man every day than a hundred every night.'

Fergus went to Cúchulainn therefore with the proposal.
He was followed by Etarcomol, son of Eda and Léthrenn,
a foster-son of Ailill and Medb.

'I would rather you didn't come,' Fergus said. 'Not that
I dislike you, but for fear of strife between Cúchulainn
and you. With your pride and insolence, and the other's
ferocity and grimness, force, fury and violence, no good
can come from your meeting.'

'Can I not be under your protection?' Etarcomol said.

'Yes,' Fergus said, 'but only if you don't insult him while he is talking.'

They went to Delga in two chariots.

It happened that Cúchulainn was playing *buanbach* with Laeg. Cúchulainn was facing away from them and Laeg facing toward them.

'I see two chariots coming,' Laeg said. 'In the first chariot there is a great dark man. His hair is dark and full. A purple cloak is wrapped about him, held by a gold brooch. He wears a red-embroidered hooded tunic. He carries a curved shield with a scalloped edge of light gold and a stabbing-spear bound around from its neck to its foot. There is a sword as big as a boat's rudder at his thigh.'

'A big empty rudder,' Cúchulainn said. 'That is my friend Fergus and it isn't a sword, but a stick, he has in his scabbard. I have heard that Ailill caught him off guard when he slept with Medb, and stole his sword and gave it to his charioteer to keep. A wooden sword was put in the scabbard.'

Fergus came up.

'Welcome, friend Fergus,' Cúchulainn said. 'If the salmon were swimming in the rivers or river-mouths I'd give you one and share another. If a flock of wild birds were to alight on the plain I'd give you one and share another; with a handful of cress or sea-herb and a handful of marshwort; and a drink out of the sand; and myself in your place in the ford of battle, watching while you slept.'

'I believe you,' Fergus said, 'but it isn't for food we came here. We know the style you keep.'

Then Cúchulainn heard Fergus's message, and Fergus left. Etarcomol stayed, staring at Cúchulainn.

'What are you staring at?' Cúchulainn said.

'You,' Etarcomol said.

'You could take that in at a glance,' Cúchulainn said.

'So I see,' Etarcomol said. 'I see nothing to be afraid of — no horror or terror or the grinding of multitudes. You're a fine lad, I imagine, for graceful tricks with wooden weapons.'

'You are making little of me,' Cúchulainn said, 'but for Fergus's sake I won't kill you. If you hadn't his protection, you would have had your bowels ripped out by now and your quarters scattered behind you all the way from your chariot to the camp.'

'You needn't threaten me any more,' Etarcomol said. 'I'll be the first of the men of Ireland to come against you tomorrow under this fine plan of single combats.'

And he went off.

He turned at Methe and Cethe, and said to his charioteer:

'I have sworn in front of Fergus,' he said, 'to fight Cúchulainn tomorrow, but I can't wait so long. Turn the horses round from this hill again.'

Laeg saw this and said to Cúchulainn:

'The chariot is coming back. He has turned the left chariot-board against us.'

'I can't refuse that,' Cúchulainn said. 'Drive down to the ford to him, and we will see.'

'It's you who want this,' Cúchulainn said to Etarcomol. 'It isn't my wish.'

'You have no choice,' Etarcomol said.

Cúchulainn cut the sod from under his feet. He fell flat, with the sod on his belly.

'Go away now,' Cúchulainn said. 'I don't want to wash my hands after you. I'd have cut you to pieces long ago but for Fergus.'

'I won't leave it like this,' Etarcomol said. 'I'll have your head, or leave you mine.'

'It will be the latter for sure.'

Cúchulainn poked at the two armpits with his sword and the clothes fell down leaving the skin untouched.

'Now clear off!' Cúchulainn said.

'No,' Etarcomol said.

Then Cúchulainn sheared off his hair with the sword-edge as neat as a razor, leaving the skin unscratched. But the fool stubbornly persisted and Cúchulainn struck down through the crown of his head and split him to the navel.

Fergus saw the chariot passing him with only one man in it and he went back in fury to Cúchulainn.

'Demon of evil,' he said, 'you have disgraced me. You must think my cudgel is very short.'

'Friend Fergus, don't rage at me,' Cúchulainn said.

> 'You ran from Ulster
>   with no sword to your fame
> and menace me
>   like a rival or foe
> I honour mighty men
>   but vain Etarcomol
> bent under my yoke
>   gave up death flowers
> stretched in my strength
>   on the chariot cushion
> sleeping or eating
>   my heroic hard hand
> never at rest
>   don't chide friend Fergus.'

And he stooped humbly while Fergus's chariot circled him three times.

'Ask Etarcomol's charioteer was I at fault,' Cúchulainn said.

'You were not, truly,' the charioteer said.

'Etarcomol swore,' Cúchulainn said, 'he wouldn't leave until he had my head or left me his own. Which would you say was easier, friend Fergus?' Cúchulainn said.

'I think it was easier to do what you did,' Fergus said. 'He was arrogant.'

Fergus pierced Etarcomol's two heels with a spancel-ring and dragged him behind his chariot to the camp. When they were travelling over rocky ground the halves of the body split apart; when it was level the halves joined again. Medb saw this.

'That is brutal treatment for the unfortunate dog,' Medb said.

'I say he was an ignorant whelp,' Fergus said, 'to pick a fight with the irresistible great Hound of Culann.'

Then they dug a grave for him; his memorial stone was planted, his name written in ogam, and his lamentation made.

Cúchulainn murdered no more that night with his sling.

'What man have you to go against Cúchulainn to-morrow?' Lugaid said.

'Maybe tomorrow we can tell,' Maine, Ailill's son, said.

'We can find no one to go against him,' Medb said. 'Ask him for a truce while we look for someone.'

He agreed to this.

'Where can we turn?' Ailill said, 'to find an opponent for Cúchulainn.'

'He has no match in Ireland,' Medb said, 'unless Cúroi mac Dáiri comes, or the warrior Nadcranntail.'

There was one of Cúroi's people in the tent.

'Cúroi won't come,' he said. 'He has done enough in sending his men here.'

'Send a message to Nadcranntail then.'

Maine Andoe, the swift one, set out to bring the news to Nadcranntail.

'For the honour of Connacht come with us.'

'I will not,' he said, 'unless they give me their daughter Finnabair.'

He went back with them and his weapons were carried in a wagon from eastern Connacht to the camp.

'You can have Finnabair,' Medb said, 'if you go against that man there.'

'I'll do it,' he said.

Lugaid went to Cúchulainn that night.

'The news is bad. Nadcranntail will be coming against you tomorrow. You'll never resist him.'

'We'll see,' Cúchulainn said.

Nadcranntail left the camp the next morning and took nine spears of holly with him, charred and sharpened. Cúchulainn was there before him in the distance catching birds, with his chariot nearby. Nadcranntail let fly a spear at Cúchulainn. Cúchulainn toyed in mid-air with the point of the spear and his bird-catching never faltered. Likewise with the other eight spears. As the ninth spear was flung the flock of birds flew away from Cúchulainn and he sped off in pursuit. Birdlike, he stepped from point to point of the flying spears in his haste not to let the birds escape. But to everyone it seemed that Cúchulainn sped in flight before Nadcranntail.

'Look at your Cúchulainn there,' Nadcranntail said. 'He has run away.'

'And why not?' Medb said. 'A true warrior came, and a timorous sprite vanished.'

Fergus and the men of Ulster were troubled by this,

and Fiacha mac Fir Febe went to Cúchulainn to protest.

'Tell him,' Fergus said, 'it was a noble stand while he showed his bravery before men. But it would be better to hide now after fleeing from one man. He shames Ulster as well as himself.'

'Who is boasting of my flight?' Cúchulainn said.

'Nadcranntail,' Fiacha said.

'What is there to boast about? The feat I did before him is nothing to be ashamed of,' Cúchulainn said. 'If he had been carrying real weapons he wouldn't be boasting now; you know I don't kill unarmed men. Let him come tomorrow,' Cúchulainn said, 'between the hill Ochaine and the sea. As early as he wishes he'll find me waiting, with no question of flight.'

So Cúchulainn went to their meeting-place and watched through the night. In the morning he flung a cloak about himself and also, without noticing it, about a great standing stone nearby, as big as himself. He came with it wrapped between his body and his cloak, and it settled upright beside him. Then Nadcranntail came, with his weapons in their wagon.

'Show me Cúchulainn,' he said.

'There he is,' Fergus said.

'He seems not quite the same as yesterday,' Nadcranntail said. 'Are you really Cúchulainn?'

'What if I am?' Cúchulainn said.

'If you are,' Nadcranntail said, 'how can I take a little lamb's head back to the camp? I can't behead a beardless boy.'

'I'm not the one,' Cúchulainn said. 'You'll find him behind that hill.'

Cúchulainn ran to Laeg.

'Make me a false beard. I can't get this warrior to fight me unless I have a beard.'

Laeg did as he asked, and Cúchulainn went to meet
Nadcranntail on the hill.

'This is more like him,' Nadcranntail said. 'A fight with
rules!'

'Agreed,' Cúchulainn said. 'Name your rules.'

'Thrown spears,' Nadcranntail said, 'and no dodging.'

'No dodging,' Cúchulainn said, 'except upward!'

Nadcranntail made a cast at him but Cúchulainn leaped
on high and it struck the standing stone and shattered in
two.

'You have fought foul! You have dodged my throw,'
Nadcranntail said.

'You are free to dodge mine by leaping upward,'
Cúchulainn said.

Then he let fly his spear, but he threw it up on high so
that it dropped down into Nadcranntail's skull and pinned
him into the earth, and Nadcranntail cried:

'Misery! Misery!'

Then he said:

'You are the best warrior in Ireland. I have twenty-
four sons in the camp. Let me go and tell them about this
treasure you've hidden in me, and I'll come back to be
beheaded. If this spear is taken out of my head I will die.'

'Agreed,' Cúchulainn said. 'But come back.'

Nadcranntail made his way back to the camp. They all
came to meet him asking:

'Where is the head of the Warped One?'

'Warriors, you will have to wait. I have things to tell
my sons. Then I go back to the fight with Cúchulainn.'

In a while he made toward Cúchulainn again and flung
his sword at him. Cúchulainn leaped on high. Then he
swelled with fury as when he faced the boy-troop in
Emain. He sprang onto the rim of Nadcranntail's shield
and struck his head off. He struck Nadcranntail again

through the neck, down to the navel, so that he fell in four sections to the ground. Then Cúchulainn chanted:

> 'Nadcranntail is no more.
> The fight grows furious.
> I could meet at this moment
> a third of Medb's men.'

VIII THE BULL IS FOUND. FURTHER SINGLE COMBATS. CUCHULAINN AND THE MORRIGAN

THEN with a third of her force, Medb set out into the district of Cuib to search for the bull, and Cúchulainn followed. It was her plan to lay waste the lands of the Ulstermen and the Picts along the Midluachair road northward as far as Dún Sobairche.

Cúchulainn caught sight of Buide mac Báin at the head of three score of Ailill's men, all in cloaks, coming from the direction of Sliab Cuilinn. They had the bull with them, surrounded by fifteen heifers. Cúchulainn went up to them.

'Where did you get these cattle?' Cúchulainn said.

'From that mountain there,' the leader said.

'Where are their herdsmen?' Cúchulainn said.

'We found only one, and we have him with us,' the warrior said.

Cúchulainn went to the ford after them in three great strides and spoke again, saying to the leader:

'What is your name?'

'One who neither fears nor favours you,' he said. 'Buide mac Báin.'

'Well, Buide, here is a spear for you!' Cúchulainn said, and he flung a short spear through his armpit, severing one of his livers in two with the spear-point. Hence comes the name Ath Buide, after him who was killed on this ford. But they got the bull into the camp.

It was said at this time that Cúchulainn would be less troublesome if his javelin could be taken from him. So Ailill's satirist, Redg, was sent to get this javelin from him.

'Give me your javelin,' the satirist said.

'I'll give you any gift but that,' Cúchulainn said.

'Other gifts I don't want,' the satirist said.

Cúchulainn struck him, for refusing what he chose to offer. Then Redg said he would take away Cúchulainn's good name unless he got the javelin. So Cúchulainn flung the javelin at him and it shot through his head.

'Now, that is a stunning gift!' the satirist cried.

So Ath Tolam Sét got its name — the Ford of the Overwhelming Gift. The copper point of the javelin came to rest at a ford further east, and so Umarrith — Where the Copper Came to Rest — is the name of that ford.

The following are those Cúchulainn killed in Cuib: Nathcoirpthe near the trees named after him, Cruithen in the ford that bears his name, the herdsmen's sons at the cairn named after them, Marc on his hillock, Meille on his hill, Badb in his tower and Boguine in his marsh.

Cúchulainn turned again toward Murtheimne Plain to defend his beloved home. It is there, as you shall hear in the proper place, that he kills the men of Cronech at Focherd when he finds them pitching camp — the ten cup-bearers and the ten warriors.

Medb turned back again from the north after spending a fortnight harassing the province. She had attacked Finnmór, wife of Celtchar mac Uthidir, and taken fifty women from her at the capture of Dún Sobairche in the territory of Dál Riada. Wherever Medb rested her horse-whip in the district of Cuib, the name Bile Medba, Medb's Whip, has remained. Any ford or height she stopped at is called Medb's Ford or Medb's Hill. They all met again at Focherd, Ailill and Medb and the troop that drove the bull. Then that herdsman who had been captured with the bull tried to make off with it, but they drove the herd after him into a narrow gap, with the beating of shafts on shields, and there the animals' hooves drove him into the earth. Forgaimen was the cowherd's name and his body is there still, giving that hill its name. They would have rested easy that night if only a man could be found to withstand Cúchulainn at the ford.

They sent for Cúr mac Daláth to fight Cúchulainn. When Cúr drew blood from a man that man died in nine days at the latest.

'If he kills Cúchulainn, we have won,' Medb said. 'But even if he is killed himself it will still take a burden off our army: it is no pleasure to be near him, sitting, sleeping or feeding.'

Cúr went forth, but he drew back when he saw a beardless boy opposing him.

'This is unfitting,' he said. 'You pay my skill a great compliment! If I knew this was the one I had to meet, I would never have come. I'll send him a boy of his own age from among my people.'

'You are mistaken,' Cormac Connlongas said, 'but it isn't surprising. We would count it a triumph if you drove him off.'

'Well, I have undertaken to do it,' Cúr said. 'But get ready to leave in the morning early. The killing of this young deer won't delay us.'

He went to meet him early next morning, having told

the armies to get ready for departure, that his meeting with
Cúchulainn would lighten the journey. It happened that
Cúchulainn was trying his special feats of arms — the
apple-feat, the feats of the sword-edge and the sloped
shield; the feats of the javelin and rope; the body-feat, the
feat of Cat and the heroic salmon-leap; the pole-throw
and the leap over a poisoned stroke; the noble chariot-
fighter's crouch; the *gae bolga;* the spurt of speed; the
feat of the chariot-wheel and the feat of the shield-rim;
the breath-feat; the snapping mouth and the hero's scream;
the stroke of precision and the stunning-shot; stepping on
a lance in flight, and straightening erect on its point; and
the trussing of a warrior.

For the first third of the day, Cúr plied his weapons on
Cúchulainn from the shelter of his shield, but couldn't
reach him with thrust or cast, Cúchulainn was so intent
on his feats. Cúchulainn didn't know that anyone was
attacking him until Fiacha mac Fir Febe said to him:

'Watch out for that man attacking you!'

Cúchulainn looked about him, and flung the one apple
left in his hand. It flew between the shield-rim and frame
and broke out through the back of the brute's head.

Fergus went back along the road toward the host.

'You are bound by your pact now,' he said, 'to wait
another day.'

'But not here,' Ailill said. 'Let us go back to our tents.'

Next, Láth mac Dabró was asked, like Cúr, to go
against Cúchulainn, and he too fell. And Fergus went
back again to remind them of the pact. So they were kept
there while Cúr mac Daláth and Láth mac Dabró were
killed, and also Foirc mac trí n-Aignech, descendant of
the three Swift Ones, and Srúbgaile mac Eobith. All were
slain in single combat.

'Go to the camp, friend Laeg,' Cúchulainn said. 'and ask Lugaid mac Nois Allchomaig who is to come against me tomorrow. Make sure you find out, and give him my greetings.'

Laeg set off.

'You are welcome,' Lugaid said. 'What a luckless man is Cúchulainn in his trouble — one man against the men of Ireland. For it is Ferbaeth goes to meet him tomorrow — may his weapons be cursed! — a comrade of Cúchulainn's and mine. They have promised him Finnabair for it, and kingship over his people.'

Laeg went back to Cúchulainn.

'My friend Laeg doesn't seem very happy with the answer,' Cúchulainn said.

Laeg told him his news.

Now Ferbaeth had been called to Ailill's and Medb's tent and told to sit by Finnabair's side. He was told that he was to have her, and that she had picked him to fight Cúchulainn. They called him their man of strength because he had had the same training with Scáthach as Cúchulainn had. They gave him wine until he was drunk, telling him it was their best out of the only fifty wagonloads they had taken with them. The girl herself handled his portion.

'I don't want all this,' Ferbaeth said. 'Cúchulainn is my foster-brother and sworn to me for ever. Still, I'll meet him tomorrow and hack his head off.'

'So you will,' Medb said.

Cúchulainn sent Laeg to ask Lugaid to come and talk with him, and Lugaid came.

'So it is Ferbaeth who is to meet me tomorrow,' Cúchulainn said.

'Yes,' Lugaid said.

'It is a black day,' Cúchulainn said. 'I won't live to see

its close. We two are of equal age and alertness. We'll meet as a perfect match. Greet him for me, friend Lugaid, and tell him it is false heroism to oppose me. Ask him to come and talk with me tonight.'

Lugaid told him this, and Ferbaeth agreed. He went that night with Fiacha mac Fir Febe to renounce his friendship with Cúchulainn. Cúchulainn begged him, by their foster-brotherhood and by their common foster-nurse, Scáthach.

'I can't,' Ferbaeth said. 'I have promised Medb.'

'Keep your friendship then!' Cúchulainn said, and left him in a fury. In the glen a piece of split holly drove into Cúchulainn's foot and its point came out at his knee. He pulled it out.

'Wait, Ferbaeth, look what I've found.'

'Throw it away,' Ferbaeth said.

Cúchulainn flung the holly-spear after Ferbaeth. It pierced the hollow at the back of his head and came out of his mouth in front, and he fell backward in the glen.

'That was a throw!' Ferbaeth said.

Some say that this is how Focherd in Murtheimne — the Place of the Throw — got its name. Others say it was Fiacha mac Fir Febe who said: 'Your throw is sharp today, Cúchulainn,' and that thus Focherd Murtheimne was named. Ferbaeth died in the glen; hence the name Glenn Firbaith, Ferbaeth's Glen.

And Fergus was heard chanting:

> 'Ferbaeth, your fool's foray
> has led to a grave in the ground.
> Your rage has brought you ruin
> and an ending in Cróen Chorann.
>
> This place in Cronech Murtheimne
> called Fichi from of old

          shall be called Focherd forever,
          where you fell, Ferbaeth.
'Your comrade is fallen,' Fergus said. 'I wonder will
you pay for his death tomorrow?'
'Sometime I must pay,' Cúchulainn said.
He sent Laeg to see what was the news in the camp, and
to find out if Ferbaeth still lived, but Lugaid said:
'Ferbaeth is dead.'
And Cúchulainn went and talked with them.

'Someone else will have to meet him tomorrow,' Lugaid
said.
    Ailill said:
'You'll get no one unless you use trickery. Give wine
to anyone who comes — it will give him courage —
and tell him: "This is the last of the wine we took
from Cruachan; we wouldn't like you to have to drink
water in our camp." Then put Finnabair at his right
hand and say: "She is yours if you bring us the head of
the Warped One." '
Each night a great warrior was called in and they made
him the offer, but each in turn was killed. At last they
could find no one to go against him. Then they turned
to Láréne mac Nois, brother to Lugaid king of Munster,
and a vainglorious man. They gave him the wine and put
Finnabair at his right hand. Medb looked at the pair.
'There is a handsome couple,' she said. 'They would
make a fine match.'
'Sure enough,' Ailill said. 'He can have her if he brings
me the head of the Warped One.'
'I'll bring it,' Láréne said.
Lugaid came up to them.
'What man have you got for the ford tomorrow?'

'Láréne,' Ailill said.

Then Lugaid went and spoke with Cúchulainn. They met at Ferbaeth's Glen and greeted each other.

'I am here to talk about Láréne, my mad boastful fool of a brother,' Lugaid said. 'They have tricked him now with the same girl. For the sake of our friendship don't kill him and leave me brotherless. He is only being sent to stir up a quarrel between us two. But I don't mind if you punish him heavily; he is coming against my wishes.'

Next day Láréne went to meet Cúchulainn, with the girl beside him urging him on. Cúchulainn sprang at him unarmed and took his weapons away roughly. He grasped him in his two hands and ground and rattled him until the dung was forced out of him. The ford grew foul with his droppings. In every direction the air thickened with his dust. Then Cúchulainn flung him into Lugaid's arms. Ever afterward, for as long as he lived, Láréne couldn't empty his bowels properly; he was never free from chest-pains; he couldn't eat without groaning. Yet he is the only man of all who met Cúchulainn on the Táin Bó Cuailnge who escaped him alive — though it was a cruel escape.

Cúchulainn beheld at this time a young woman of noble figure coming toward him, wrapped in garments of many colours.

'Who are you?' he said.

'I am King Buan's daughter,' she said, 'and I have brought you my treasure and cattle. I love you because of the great tales I have heard.'

'You come at a bad time. We no longer flourish here, but famish. I can't attend to a woman during a struggle like this.'

'But I might be a help.'

'It wasn't for a woman's backside I took on this ordeal!'

'Then I'll hinder,' she said. 'When you are busiest in the fight I'll come against you. I'll get under your feet in the shape of an eel and trip you in the ford.'

'That is easier to believe. You are no king's daughter. But I'll catch and crack your eel's ribs with my toes and you'll carry that mark forever unless I lift it from you with a blessing.'

'I'll come in the shape of a grey she-wolf, to stampede the beasts into the ford against you.'

'Then I'll hurl a sling-stone at you and burst the eye in your head, and you'll carry that mark forever unless I lift it from you with a blessing.'

'I'll come before you in the shape of a hornless red heifer and lead the cattle-herd to trample you in the waters, by ford and pool, and you won't know me.'

'Then I'll hurl a stone at you,' he said, 'and shatter your leg, and you'll carry that mark forever unless I lift it from you with a blessing.'

Then she left him.

Lóch mac Mofemis was asked next. They promised him a part of the fine Plain of Ai equal to the Plain of Murtheimne, with war-harness for a dozen men and a chariot worth seven bondmaids. But he thought it beneath him to fight with a boy. He had a brother Long mac Mofemis, and to him in turn they offered the same reward: the girl, the war-harness, the chariot and the land. He fought Cúchulainn and Cúchulainn slew him and he was carried in death up to his brother Lóch. Lóch said that if he could be sure it was a grown man that had killed him, he would kill him for it.

The women called out to Cúchulainn that people in the camp were mocking at him because he had no beard; that it was only reckless men, and not their best warriors, that would fight him; and that it would be better if he made a beard with berry juice. He did this, to get Lóch to fight him. And he plucked a fistful of grass and spoke into it and everyone believed he had a beard.

'Look,' the women said, 'Cúchulainn is bearded. A warrior may fight him now.'

They did this to urge on Lóch, but Lóch said:

'I won't fight him for seven days.'

'We can't leave him in peace for so long,' Medb said. 'Send out a warrior every night to steal up and catch him off guard.'

They did this; a warrior stole out to find him each night, but he killed them all. These are the names of the men who fell there: seven who were named Conall, seven named Aengus, seven named Uargus, seven named Celtre, eight named Fiac, ten named Ailill, ten named Delbath and ten named Tasach. Those were his week's deeds at Ath Grencha.

Then Medb began to incite Lóch.

'It is a great shame on you,' she said, 'that the man who killed your brother can destroy our army, and you still haven't gone to fight him. Surely a peppery overgrown elf like him can't resist the fiery force of a warrior like you. Wasn't it from the same teacher and foster-mother you both learned your skill?'

Lóch went out to meet him and avenge his brother, satisfied that he was going to meet a bearded man.

'Come to the ford upstream,' Lóch said. 'I won't meet you in this foul place where Long fell.'

While Cúchulainn was going to that ford men drove some cattle over.

'There will be a great trampling across your water here today,' Gabrán the poet said.

Ath Tarteisc, 'across your water,' and Tír Mór Tairtesc, the mainland of Tarteisc, got their names in this way.

The men met there in the ford and fought and struck at each other. As they were exchanging blows an eel flung three coils about Cúchulainn's feet and he fell back in the ford. Then Lóch set upon Cúchulainn with the sword until the ford was blood-red with his crimson gore.

'Urge him on!' Fergus said to his followers. 'This is a poor spectacle in front of the enemy. Let someone put heart in Cúchulainn or he will die for want of encouragement.'

The venom-tongued Bricriu mac Carbad stood up and started to taunt Cúchulainn.

'Your strength is withered up,' Bricriu said, 'if a little salmon can put you down like this, and the men of Ulster rising out of their pangs. If this is what happens when you meet a tough warrior in arms, it's a pity you took on a hero's task, with all the men of Ireland looking on.'

Cúchulainn rose up at this and struck the eel and smashed its ribs. Then, with the thunderous deeds that the warriors did in the ford, the cattle stampeded madly eastward through the army and carried off the tents on their horns. Next a she-wolf attacked Cúchulainn and drove the cattle back westward upon him, but he let fly a stone from his sling and burst the eye in her head. She came in the shape of a hornless red heifer and led the cattle dashing through the fords and pools, so that he cried out:

'I can't tell ford from flood!'

He slung a stone at the hornless red heifer and broke her legs beneath her. So it was that Cúchulainn did to the

Morrígan the three things he had sworn. He made this
chant:

> 'I am alone against hordes.
> I can neither halt nor let pass.
> I watch through the long hours
> alone against all men.
>
> Tell Conchobor to come now.
> It wouldn't be too soon.
> Mágach's sons have stolen our cattle
> to divide between them.
>
> I have held them single-handed,
> but one stick won't make fire.
> Give me two or three
> and torches will blaze!
>
> I am almost worn out
> by single contests.
> I can't kill all their best
> alone as I am.'

Then he fought Lóch with the sword and the *gae bolga*
that his charioteer sent him along the stream. He struck
him with it up through the fundament of his body — for
when Lóch was fighting, all his other parts were covered
in a skin of horn.

'Yield to me: leave me space,' Lóch said.

Cúchulainn yielded before him and Lóch fell forward
on his face. From this Ath Traigid is named in Tír Mór —
the Ford of Yielding. Then Cúchulainn cut his head off.

A great weariness fell on Cúchulainn. The Morrígan
appeared to him in the shape of a squint-eyed old woman

milking a cow with three teats. He asked her for a drink and she gave him milk from the first teat.

'Good health to the giver!' Cúchulainn said. 'The blessing of God and man on you.'

And her head was healed and made whole. She gave him milk from the second teat and her eye was made whole. She gave him milk from the third teat and her legs were made whole.

'You said you would never heal me,' the Morrígan said.

'If I had known it was you I wouldn't have done it,' Cúchulainn said.

IX   THE PACT IS BROKEN: THE GREAT CARNAGE

ASK Cúchulainn for a truce,' Ailill and Medb said.

Lugaid went to ask him and Cúchulainn granted the truce.

'But have a man at the ford for me tomorrow,' Cúchulainn said.

Now there were six paid soldiers of royal blood in Medb's army, six sons of kings of the Clanna Dedad. They were known as the three Dark-haired Ones of Imlech and the three Red-heads of Sruthar.

'Why shouldn't we go all together against Cúchulainn?' they said.

So they went against him on the next day, and Cúchulainn slew all six.

Medb considered again what to do with Cúchulainn. She was greatly troubled by the number being killed in her army. She decided to ask him to meet her and talk with her at a certain place, and then set a great number of keen and spirited men on him. So she sent her messenger, with a false offer of peace, to find Cúchulainn and get him to meet her at that place next day. He was to come unarmed and she was to go by herself, with only her troop of women in attendance. Traigthrén was the messenger — the strong of foot — and he went up to Cúchulainn and gave him Medb's message. Cúchulainn said he would do what she asked.

'Cúchulainn,' Laeg said, 'how do you plan to go to this meeting with Medb tomorrow?'

'The way Medb asked me,' Cúchulainn said.

'Medb is a forceful woman,' the charioteer said. 'I'd watch out for her hand at my back.'

'How should I go?' Cúchulainn said.

'With your sword at your side,' the charioteer said, 'not to be caught off guard. A warrior without his weapons is not under warriors' law; he is treated under the rule for cowards.'

'I'll do as you say,' Cúchulainn said.

The meeting was fixed for the hill Ard Aighnech, called Focherd today. Medb came there and set a trap for Cúchulainn with fourteen of her own most skillful follow-ers: two named Glas Sinna, two sons of Buccride, two named Ardán, two sons of Lecc, two named Glas Ogma, two sons of Cronn, with Drucht and Delt and Daithen, Tea and Tascur and Tualang, Taur and Glese.

When Cúchulainn came to the meeting place the men rose up against him. Fourteen javelins were hurled at him together but Cúchulainn guarded himself so that his skin was untouched, and even his armour. Then he turned on them and killed all fourteen of them. These are the 'Fourteen at Focherd,' who are also remembered as 'the warriors of Cronech,' for it was in Cronech near Focherd that they died.

Then Cúchulainn chanted:

> 'My skill in arms grows great.
> On fine armies cowering
> I let fall famous blows.
> On whole hosts I wage war
> to crush their chief hero
> and Medb and Ailill also
> who stir up wrong, red hatred
> and black woman-wailing,
> who march in cruel treachery
> trampling their chief hero
> and his sage, sound advice
> — a fierce, right-speaking warrior
> full of noble acts.'

Some believe that the name Focherd comes from the opening words of this chant, 'Fo . . . cherd'— the 'great skill' of Cúchulainn there.

Then Cúchulainn fell upon the army as they were settling their camp, and killed two named Dagrí and two named Anle and four named Dúngas from Imlech. Afterward on the same day they again fought him foul. Five went out against him together — two named Cruaid, two named Calad, and Derothor—and Cúchulainn killed them single-handed.

Fergus said they must stop breaking the rule of fair fight against Cúchulainn, and Cúchulainn did single combat until they reached Delga in Murtheimne — at that time called Dún Cinn Coros. Cúchulainn killed Fota in the field now called by his name, Bómailce on his ford, Salach in his marsh, Muinne on his hill, Luar in Lethbera Luair, and Fertóithle in Tóithli. So these places are named

forever after the men who fell there. Cúchulainn slew
Traig and Dornu and Dernu — Foot, Fist and Palm — and
Col, Mebul and Eraise — Lust, Shame and Nothingness
— on the near side of the ford of Tír Mór at Methe and
Cethe; these were three druids and their three wives.
After this Medb sent out one hundred of her own follow-
ers to kill Cúchulainn but he slew them all at the ford of
Cét Chuile — the Crime of One Hundred. For it was here
that Medb said: 'A crime, this slaughter of our people!'
From this episode came the names Glais Chrau — the
Stream of Blood; Cuilenn Cinn Dúin — the Crime (some
say) of Cinn Dúin, the head of the fort; and the Ford of
Cét Chuile.

Then he pelted them from where he was in Delga so
that no living thing, man or beast, dared show its face
past him southward between Delga and the sea.

'Take this message to him,' Ailill said: 'he can have
Finnabair if he leaves our armies alone.'

Lugaid went and told Cúchulainn about this offer.

'Friend Lugaid,' Cúchulainn said, 'I don't trust them.'

'It is the word of a king,' Lugaid said. 'It is no lie.'

'I accept, so,' Cúchulainn said.

Lugaid brought Cúchulainn's answer back to Ailill and
Medb.

'Send the camp fool made up to look like me,' Ailill
said, 'with a king's crown on his head. Stand him at a
distance from Cúchulainn so as not to be recognised, and
send the girl with him. He can betroth her to Cúchulainn
and they can come away quickly. Maybe the trick will
work and hold him back until the day when he comes
with the men of Ulster to the last Battle.'

The fool Tamun — the Stump — was sent with the

girl and he spoke from a distance to Cúchulainn.
Cúchulainn went to meet them and knew by the man's
speech that he was the camp fool. He shot a sling-stone
from his hand and pierced the fool's head and knocked
out his brains. Cúchulainn went up to the girl and cut off
her two long tresses and thrust a pillar-stone under her
cloak and tunic. He thrust another pillar-stone up through
the fool's middle. Their two standing-stones are there still,
Finnabair's Pillar-Stone and the Fool's Pillar-Stone. Cúchulainn left them like that. Some of Ailill and Medb's
people came looking for them because they stayed away
so long, and saw their condition, and the story spread
through the whole encampment. There was no further
truce for them with Cúchulainn after that.

The four provinces of Ireland settled down and camped
on Murtheimne Plain, at Breslech Mór (the place of their
great carnage). They sent their shares of cattle and plunder
southward ahead of them to Clithar Bó Ulad, the Cattle-
Shelter of Ulster. Cúchulainn took his place near them
at the gravemound in Lerga. At nightfall his charioteer
Laeg mac Riangabra kindled a fire for him. And he saw
in the distance over the heads of the four provinces of
Ireland the fiery flickering of gold weapons in the evening
sunset clouds. Rage and fury seized him at the sight of
that army, at the great forces of his foes, the immensity
of his enemies. He grasped his two spears, his shield and
his sword and he shook the shield and rattled the spears
and flourished the sword and gave the warrior's scream
from his throat, so that demons and devils and goblins
of the glen and fiends of the air replied, so hideous
was the call he uttered on high. Then the Nemain stirred
the armies to confusion. The weapons and spear-points of

the four armed provinces of Ireland shook with panic.
One hundred warriors fell dead of fright and terror that
night in the heart of the guarded camp.

Laeg stood in his place and saw a solitary man crossing
between the camp of the men of Ireland straight toward
him out of the northeast.

'There is a man coming toward us alone, Little Hound,'
Laeg said.

'What kind of man is he?', Cúchulainn said.

'It is soon told: a tall, broad, fair-seeming man. His
close-cropped hair is blond and curled. A green cloak is
wrapped about him, held at his breast by a bright silver
brooch. He wears a knee-length tunic of kingly silk, red-
embroidered in red gold, girded against his white skin.
There is a knob of light gold on his black shield. He carries
a five-pointed spear in his hand and a forked javelin. His
feats and graceful displays are astonishing, yet no one is
taking any notice of him and he heeds no one: it is as
though they couldn't see him.'

'They can't, my young friend,' Cúchulainn said. 'This
is some friendly one of the *síde* that has taken pity on me.
They know my great distress now on the Táin Bó Cuailnge,
alone against all four provinces of Ireland.'

Cúchulainn was right. When the warrior came up to
him he said in pity:

'This is a manly stand, Cúchulainn.'

'It isn't very much,' Cúchulainn said.

'I am going to help you now,' the warrior said.

'Who are you?' Cúchulainn said.

'I am Lug mac Ethnenn, your father from the *síde*.'

'My wounds are heavy. It is time they were let heal.'

'Sleep a while, then, Cúchulainn,' the warrior said, 'a

heavy sleep of three days and three nights by the grave-mound at Lerga. I'll stand against the armies for that time.'

He sang to Cúchulainn, as men sing to men, until he slept. Then he examined each wound and cleaned it. Lug made this chant:

'Rise son of mighty Ulster
　　　with your wounds made whole
a fair man faces your foes
　　　in the long night over the ford
rest in his human care
　　　everywhere hosts hewn down
succour has come from the *síde*
　　　to save you in this place
your vigil on the hound fords
　　　a boy left on lonely guard
defending cattle and doom
　　　kill phantoms while I kill
they have none to match your span
　　　of force or fiery wrath
your force with the deadly foe
　　　when chariots travel the valleys
then arise arise my son.'

Cúchulainn slept three days and three nights, and well he might; for if his sleep was deep so was his weariness. From the Monday after the feast of Samain at summer's end to the Wednesday after the feast of Imbolc at spring's beginning, Cúchulainn never slept — unless against his spear for an instant after the middle of the day, with head on fist and fist on spear and the spear against his knee — for hacking and hewing and smiting and slaughtering the four great provinces of Ireland.

Then the warrior from the *síde* dropped wholesome healing herbs and grasses into Cúchulainn's aching wounds and several sores, so that he began to recover in his sleep without knowing it.

The boy-troop in Ulster spoke among themselves at this time.

'It is terrible,' they said, 'that our friend Cúchulainn must do without help.'

'Let us choose a company to help him,' Fiachna Fuilech, the Bloodspiller, said — a brother of Fiacha Fialdána mac Fir Febe.

Then the boy-troop came down from Emain Macha in the north carrying their hurling-sticks, three times fifty sons of Ulster kings — a third of their whole troop — led by Follamain, Conchobor's son. The army saw them coming over the plain.

'There is a great number crossing the plain toward us,' Ailill said.

Fergus went to look.

'These are some of the boy-troop of Ulster coming to help Cúchulainn,' he said.

'Send out a company against them,' Ailill said, 'before Cúchulainn sees them. If they join up with him you'll never stand against them.'

Three times fifty warriors went out to meet them, and they all fell at one another's hands at Lia Toll, the Pierced Standing-Stone. Not a soul came out alive of all those choice children except Follamain mac Conchoboir. Follamain swore he would never go back to Emain while he drew breath, unless he took Ailill's head with him, with the gold crown on top. But that was no easy thing to swear; the two sons of Bethe mac Báin, sons of Ailill's foster-mother and foster-father, went out and attacked him, and he died at their hands.

'Make haste,' Ailill said, 'and ask Cúchulainn to let you move on from here. There will be no forcing past him once his hero-halo springs up.'

Cúchulainn, meanwhile, was sunk in his sleep of three days and nights by the gravemound at Lerga. When it was done he rose up and passed his hand over his face and turned crimson from head to foot with whirling excitement. His spirit was strong in him; he felt fit for a festival, or for marching or mating, or for an ale-house or the mightiest assembly in Ireland.

'Warrior!' Cúchulainn said. 'How long have I been in this sleep?'

'Three days and three nights,' the warrior said.

'Alas for that!' Cúchulainn said.

'Why?' the warrior said.

'Because their armies were free from attack all that time,' Cúchulainn said.

'They were not,' the warrior said.

'Tell me what happened,' Cúchulainn said.

'The boy-troop came south from Emain Macha, three times fifty sons of Ulster kings, led by Follamain, Conchobor's son, and they fought three battles with the armies in the three days and nights you slept, and they slew three times their own number. All the boy-troop perished except Follamain mac Conchoboir. Follamain swore to take home Ailill's head, but that was no easy thing, and he too was killed.'

'Shame,' Cúchulainn said, 'that I hadn't my strength for this! If I had, the boy-troop wouldn't have perished as they did and Follamain mac Conchoboir wouldn't have fallen.'

'Onward, Little Hound; there is no stain on your good name, no slight on your courage.'

'Stay with us tonight,' Cúchulainn said, 'and we'll

avenge the boy-troop together.'

'I will not stay,' the warrior said. 'No matter what deeds of craft or courage a man does in your company the glory and fame and name go to you, not to him. So I will not stay. Go bravely against the army by yourself. They have no power over your life at this time.'

'The sickle chariot, friend Laeg,' Cúchulainn said, 'can you yoke it? Have you everything needed? If you have, get it ready. If you haven't, leave it be.'

The charioteer rose up then and donned his charioteer's war-harness. This war-harness that he wore was: a skin-soft tunic of stitched deer's leather, light as a breath, kneaded supple and smooth not to hinder his free arm movements. He put on over this his feathery outer mantle, made (some say) by Simon Magus for Darius king of the Romans, and given by Darius to Conchobor, and by Conchobor to Cúchulainn, and by Cúchulainn to his charioteer. Then the charioteer set down on his shoulders his plated, four-pointed, crested battle-cap, rich in colour and shape; it suited him well and was no burden. To set him apart from his master, he placed the charioteer's sign on his brow with his hand: a circle of deep yellow like a single red-gold strip of burning gold shaped on an anvil's edge. He took the long horse-spancel and the ornamental goad in his right hand. In his left hand he grasped the steed-ruling reins that give the charioteer control. Then he threw the decorated iron armour-plate over the horses, covering them from head to foot with spears and spit-points, blades and barbs. Every inch of the chariot bristled. Every angle and corner, front and rear, was a tearing-place.

He cast a protecting spell on his horses and his

companion-in-arms and made them obscure to all in the
camp, while everything remained clear to themselves.
It was well he cast such a spell, for he was to need
his three greatest charioteering skills that day : leaping
a gap, straight steering and the use of the goad.

Then the high hero Cúchulainn, Sualdam's son, builder
of the Badb's fold with walls of human bodies, seized his
warrior's battle-harness. This was the warlike battle-
harness he wore : twenty-seven tunics of waxed skin,
plated and pressed together, and fastened with strings
and cords and straps against his clear skin, so that his
senses or his brain wouldn't burst their bonds at the onset
of his fury. Over them he put on his heroic deep battle-
belt of stiff, tough, tanned leather from the choicest parts
of the hides of seven yearlings, covering him from his
narrow waist to the thickness of his armpit; this he wore
to repel spears or spikes, javelins, lances or arrows — they
fell from it as though dashed at stone or horn or hard
rock. Then he drew his silk-smooth apron, with its light-
gold speckled border, up to the softness of his belly. Over
this silky skin-like apron he put on a dark apron of well-
softened black leather from the choicest parts of the hides
of four yearlings, with a battle-belt of cowhide to hold it.
Then the kingly champion gripped his warlike battle-
weapons. These were the warlike weapons he chose :
eight short swords with his flashing, ivory-hilted sword;
eight small spears with his five-pronged spear, and a
quiver also; eight light javelins with his ivory javelin;
eight small darts with his feat-playing dart, the *del chliss*;
eight feat-playing shields with his dark-red curved shield
that could hold a prize boar in its hollow, its whole rim
so razor sharp it could sever a single hair against the
stream. When Cúchulainn did the feat of the shield-rim he
could shear with his shield as sharply as spear or sword.

He placed on his head his warlike, crested battle-helmet,
from whose every nook and cranny his longdrawn scream
re-echoed like the screams of a hundred warriors; so it
was that the demons and devils and goblins of the glen
and fiends of the air cried out from that helmet, before
him, above him and around him, whenever he went out to
spill the blood of warriors and heroes. His concealing
cloak was spread about him, made of cloth from Tír
Tairngire, the Land of Promise. It was given to him by
his magical foster-father.

The first warp-spasm seized Cúchulainn, and made him
into a monstrous thing, hideous and shapeless, unheard of.
His shanks and his joints, every knuckle and angle and
organ from head to foot, shook like a tree in the flood or
a reed in the stream. His body made a furious twist inside
his skin, so that his feet and shins and knees switched to
the rear and his heels and calves switched to the front.
The balled sinews of his calves switched to the front of
his shins, each big knot the size of a warrior's bunched
fist. On his head the temple-sinews stretched to the nape
of his neck, each mighty, immense, measureless knob as
big as the head of a month-old child. His face and features
became a red bowl: he sucked one eye so deep into his
head that a wild crane couldn't probe it onto his cheek
out of the depths of his skull; the other eye fell out along
his cheek. His mouth weirdly distorted: his cheek peeled
back from his jaws until the gullet appeared, his lungs and
liver flapped in his mouth and throat, his lower jaw
struck the upper a lion-killing blow, and fiery flakes large
as a ram's fleece reached his mouth from his throat. His
heart boomed loud in his breast like the baying of a
watch-dog at its feed or the sound of a lion among bears.
Malignant mists and spurts of fire — the torches of the
Badb — flickered red in the vaporous clouds that rose

boiling above his head, so fierce was his fury. The hair
of his head twisted like the tangle of a red thornbush
stuck in a gap; if a royal apple tree with all its kingly
fruit were shaken above him, scarce an apple would reach
the ground but each would be spiked on a bristle of his
hair as it stood up on his scalp with rage. The hero-halo
rose out of his brow, long and broad as a warrior's whet-
stone, long as a snout, and he went mad rattling his
shields, urging on his charioteer and harassing the hosts.
Then, tall and thick, steady and strong, high as the mast
of a noble ship, rose up from the dead centre of his skull
a straight spout of black blood darkly and magically
smoking like the smoke from a royal hostel when a king
is coming to be cared for at the close of a winter day.

When that spasm had run through the high hero
Cúchulainn he stepped into his sickle war-chariot that
bristled with points of iron and narrow blades, with hooks
and hard prongs and heroic frontal spikes, with ripping
instruments and tearing nails on its shafts and straps and
loops and cords. The body of the chariot was spare and
slight and erect, fitted for the feats of a champion, with
space for a lordly warrior's eight weapons, speedy as the
wind or as a swallow or a deer darting over the level
plain. The chariot was settled down on two fast steeds,
wild and wicked, neat-headed and narrow-bodied, with
slender quarters and roan breast, firm in hoof and harness
— a notable sight in the trim chariot-shafts. One horse
was lithe and swift-leaping, high-arched and powerful,
long-bodied and with great hooves. The other flowing-
maned and shining, slight and slender in hoof and heel.

In that style, then, he drove out to find his enemies and
did his thunder-feat and killed a hundred, then two
hundred, then three hundred, then four hundred, then five
hundred, where he stopped — he didn't think it too many

to kill in that first attack, his first full battle with the
provinces of Ireland. Then he circled the outer lines of
the four great provinces of Ireland in his chariot and he
attacked them in hatred. He had the chariot driven so
heavily that its iron wheels sank into the earth. So deeply
the chariot-wheels sank in the earth that clods and
boulders were torn up, with rocks and flagstones and the
gravel of the ground, in a dyke as high as the iron wheels,
enough for a fortress-wall. He threw up this circle of the
Badb round about the four great provinces of Ireland to
stop them fleeing and scattering from him, and corner
them where he could wreak vengeance for the boy-troop.
He went into the middle of them and beyond, and
mowed down great ramparts of his enemies' corpses,
circling completely around the armies three times, attacking them in hatred. They fell sole to sole and neck to
headless neck, so dense was that destruction. He circled
them three times more in the same way, and left a bed
of them six deep in a great circuit, the soles of three to
the necks of three in a ring around the camp. This
slaughter on the Táin was given the name Seisrech Bresligi,
the Sixfold Slaughter. It is one of the three uncountable
slaughters on the Táin: Seisrech Bresligi, Imslige Glenn-
amnach — the mutual slaughter at Glenn Domain — and
the Great Battle at Gáirech and Irgairech (though this
time it was horses and dogs as well as men.) Any count
or estimate of the number of the rabble who fell there is
unknown, and unknowable. Only the chiefs have been
counted. The following are the names of these nobles
and chiefs: two called Cruaid, two named Calad, two
named Cír, two named Cíar, two named Ecell, three
named Crom, three named Caur, three named Combirge,
four named Feochar, four named Furechar, four named
Cass, four named Fota, five named Aurith, five named

Cerman, five named Cobthach, six named Saxan, six named Dach, six named Dáire, seven named Rochad, seven named Ronan, seven named Rurthech, eight named Rochlad, eight named Rochtad, eight named Rinnach, eight named Coirpre, eight named Mulach, nine named Daithi, nine more named Dáire, nine named Damach, ten named Fiac, ten named Fiacha and ten named Feidlimid. In this great Carnage on Murtheimne Plain Cúchulainn slew one hundred and thirty kings, as well as an uncountable horde of dogs and horses, women and boys and children and rabble of all kinds. Not one man in three escaped without his thighbone or his head or his eye being smashed, or without some blemish for the rest of his life. And when the battle was over Cúchulainn left without a scratch or a stain on himself, his helper or either of his horses.

X   COMBAT WITH FERGUS AND OTHERS

CUCHULAINN came out the next morning to view the armies and display his noble fine figure to the matrons and virgins and young girls and poets and bards. He came out to display himself by day because he felt the unearthly shape he had shown them the night before had not done him justice. And certainly the youth Cúchulainn mac Sualdaim was handsome as he came to show his form to the armies. You would think he had three distinct heads of hair — brown at the base, blood-red in the middle, and a crown of golden yellow. This hair was settled strikingly into three coils on the cleft at the back of his head. Each long loose-flowing strand hung down in shining splendour over his shoulders, deep-gold and beautiful and fine as a thread of gold. A hundred neat

red-gold curls shone darkly on his neck, and his head was
covered with a hundred crimson threads matted with
gems. He had four dimples in each cheek — yellow,
green, crimson and blue — and seven bright pupils, eye-
jewels, in each kingly eye. Each foot had seven toes and
each hand seven fingers, the nails with the grip of a
hawk's claw or a gryphon's clench. He wore his festive
raiment that day. This is what he wore: a fitted purple
mantle, fringed and fine, folded five times and held at his
white clear breast by a brooch of light-gold and silver
decorated with gold inlays — a shining source of light
too bright in its blinding brilliance for men to look
at. A fretted silk tunic covered him down to the top of
his warrior's apron of dark-red royal silk. He carried a
dark deep-red crimson shield — five disks within a light-
gold rim — and a gold-hilted sword in a high clasp on his
belt, its ivory guard decorated with gold. Near him in
the chariot he had a tall grey-bladed javelin with a hard
hungry point, rivetted with bright gold. He held in one
hand nine human heads and in the other hand ten, and
he shook them at the armies — the crop of one night's
warfare on the four provinces of Ireland.

The Connacht women climbed on the soldiers, and the
Munster women climbed on their own men, to see
Cúchulainn. But Medb couldn't see him, not daring to
show her face from under the barrel-shaped shelter of
shields for dread of him.

'This is a nuisance,' she said. 'I can't see the boy they
are making so much of.'

'It would give you no peace to see him,' Léthren, Ailill's
groom said.

'Fergus,' Medb said, 'what kind of man is he?'

'A boy who checks
    sword with shield
for cattle and women
    who makes a division
of men's bodies
    hacked and hacking
in Ulster's fords
    and sweetly shares
the royal spoils
    a fierce young man
if he is the Hound
    who calls Murtheimne
Plain his own.'

Then Medb herself climbed up on the mens' backs to see
him.

Dubthach the Dark, of Ulster, said at this time :

'Is this the Warped One?
We'll have corpses,
shrieks in our enclosures,
tales to tell,

stones over graves,
dead kings increasing.
You may battle this Brave One
but you are lost.

His wild shape I see,
and his heap of plunder —
nine heads in one hand,
and ten more, his treasure.

Your women climb up
and show their faces,
but your great queen shuns
the bitter battle.

If I had my way
all the armies together
would put an end
to this Warped One!'

But Fergus answered:

'Get Dubthach with his black tongue
back behind our army!
Since the maiden-massacre
he has done only harm.

It was base slaughter when he slew
Conchobor's son Fiacha,
and no better when he killed
Coirpre mac Feidlimid.

Now this son of Lugaid lags
in the battle against Ulster.
Those that he can't kill
he sets at each other's throats.

All the exiles would lament
the slaughter of our beardless son.
But soon the Ulster hosts will come
and harass you like herds of cattle
— your councils scattered far and wide
by Ulster risen from its pangs.

There'll be stories of great slaughter
and the crying of great queens.
There'll be mangling of wounds
and mounds made of the slain.

There'll be corpses under foot
and there'll be ravens at their meat
and shields scattered on the slopes
and sorrowing and pillaging

and blood of men in multitudes
poured out over the ground.
We have wandered far indeed
in exile from our Ulster home.'

Then Fergus flung Dubthach from him and he fell motionless near a group of soldiers.

Ailill said:
    'Fergus why so fierce
        over Ulster cows and women
    I can sense great slaughter
        and gaps of butchery
    though it is one by one
        they die in the ford each day.'
Medb said:
    'Rise up Ailill
        with triple ranks
    guard your cattle
        the grinning boy
    storms in turmoil
        by brinks of fords
    wide gravel beds
        and dark pools

valorous Fergus
> and Ulster's exiles
will have their due
> when the battle is done
with grief to crush
> poets of war.'

Fergus said:
> 'Pay no heed
>> to stupid women
> flames flowering
>> kith and kin
> buried away
>> and dire deaths
> cool your fever
>> fight fair.'

The poet Gabrán said:
> 'Why make a show of words
>> for queens and followers
> to have a taste of fierceness
>> when it comes to blades in battle
> there's one you must pin down
>> or he will earn our hatred.'

'Don't flinch from him,' Fergus said. 'Go and meet him
in the ford.'
'Let us hear from Ailill,' Medb said.
Ailill said:
> 'Fergus knows this land
>> he brings shame on your heads
> he won't lead your cattle round
>> but hacks and plunders round them
> and swears he doesn't take us
>> by ways many and long.'

Fergus said:
> 'After a year of women
> strife and bitter ending
> most noble Medb don't blame
> your exile troops too harshly
> or make little of the man
> who came to your support.'

Fiacha Fialdána, the bold and true, went to speak with his cousin Maine Andoe, the swift. Dóchae mac Mágach came with Maine Andoe, and Dubthach the Black from Ulster came with Fiacha Fialdána. Dóchae threw a javelin at Fiacha but hit his own friend Dubthach. Fiacha Fialdána threw a javelin at Dóchae but hit his kinsman Maine. The place where this happened is called Imroll Belaig Eóin, the Miscast at Bird Pass.

Some say that Imroll Belaig Eóin got its name later, when the Ulstermen had risen from their pangs: that the two armies had arrived and settled at Belach Euin when Diarmait, Conchobor's son, came south out of Ulster. He said:

'Send out a horseman. If Maine comes I'll go to meet him and the two of us can talk.'

So they met.

'I come from Conchobor,' Diarmait said, 'to tell Medb and Ailill to set the cattle free and make good all the damage they have done to us. Let them bring their bull to the bull here in the east and have them fight, as Medb promised.'

'I'll go and tell them,' Maine said, and he told them.

'Medb won't have it,' Maine said when he came back.

'We can exchange weapons then, if you like,' Diarmait said.

'I don't mind,' Maine said.

Each of them threw his javelin at the other and they killed each other. Some say that this is how Imroll Belaig Eóin got its name, and that the armies then rushed upon each other, killing three score out of each force, and that the name Ard in Dírma comes from this — the Height where the Armies were.

A brave Ulster warrior, Aengus, son of Aenlám Gaibe, turned the whole army aside at Muid Loga — the place called Lugmod today — as far as Ath Da Ferta, the Ford of the Two Grave Mounds. He refused to let them pass, and pelted them with flagstones. Some say that if they had agreed to single combat the whole army would have fallen at his hands before they came under the sword at Emain Macha. But there was no rule of fair play for him and he died overwhelmed.

'Send someone out to me,' Cúchulainn said at Ath Da Ferta.

'Not I, not I!' they all called out from their places. 'My family owes no sacrifice! And if it did, why should I be the one?'

Fergus mac Roich was asked to fight him, but he wouldn't fight his foster-son Cúchulainn. Wine was brought and he grew very drunk. He was asked again to go into combat and this time he went, because they implored him.

'You must be under strong protection, friend Fergus,' Cúchulainn said, 'to come against me with no sword in your scabbard.'

'It would be all the same if I had a sword in it,' Fergus

said, 'I wouldn't use it on you. Yield to me now, Cúchulainn,' he said.

'If you will yield to me another time,' Cúchulainn said.

'Agreed,' Fergus said.

Cúchulainn retreated back before Fergus as far as the
swamp of Grellach Dollaid on condition that Fergus
would give way to him on the day of the great Battle.
Cúchulainn ran off into Grellach Dollaid.

'Chase him, Fergus!' they all cried.

'I won't,' Fergus said. 'It isn't so easy. That one there
is too lively. I'm not going after him until my turn comes
round again.'

Then they all went past Cúchulainn and set up camp in
Crích Rois.

Ferchu Loingsech was a Connachtman, who was always
harassing and hounding Ailill and Medb. From the day
they took the kingship he had never once gone to visit
their camp, even when he was in difficulty or dire straits.
He was forever pillaging and plundering their border lands
when they were away. It happened that he was eastward
of Ai Plain with his troop of a dozen men at this time
and he heard that four of the provinces of Ireland had
been stopped and held from the Monday at summer's
end to the beginning of spring by one man, who killed
a man on the ford each day and a hundred men at night.
Ferchu discussed this with his people and said:

'The best thing we can do is attack this man who has
stopped and held the four provinces of Ireland, and bring
his head and weapons with us back to Ailill and Medb.
No matter what crimes we have committed against them,
they'll forgive us if we kill this man.'

They all agreed to this and went to find Cúchulainn.

When they found him they fought foul and fell on him all twelve together. But Cúchulainn turned on them and instantly struck off their twelve heads. He planted twelve stones for them in the ground and set a head on each stone, and Ferchu Loingsech's head on its stone as well. It is from this, where Ferchu left his head, that the name Cinnit Ferchon Loingsig comes — reading it 'Cenn áit' Ferchon, the Place of Ferchu's Head.

The next day Medb sent twenty-nine men out together against Cúchulainn in the swamp of Fuiliarnn — Blood-iron — a swamp on the near side of Ferdia's Ford. They were Gaile Dána and his twenty-seven sons, with his sister's son, Glas mac Delga. The arrangement to do this was made in Fergus's presence, and he had to agree with it. They argued that it should be counted a single combat, because the sons of Gaile Dána were all the issue of his body, limb of his limb and flesh of his flesh. Fergus went aside to his tent with his followers and uttered a tired sigh aloud.

'A sad thing is going to happen here tomorrow,' he said.

'What is that?' his followers asked.

'The killing of Cúchulainn,' Fergus said.

'Alas!' they said. 'But who can kill him?'

'Gaile Dána,' he said, 'and his twenty-seven sons and his sister's son Glas mac Delga. Every one of them has poison on him, and there is poison on all their weapons. Any man that they wound will die in nine days at most, if he doesn't die at once. If anyone will go for me to see this fight and bring me the story of Cúchulainn's death he can have my weapons and my blessing.'

'I will go,' Fiacha mac Fir Febe said.

They stayed there that night. Gaile Dána rose up early

in the morning with his twenty-seven sons and his sister's
son Glas mac Delga and they went out to find Cúchulainn.
Fiacha mac Fir Febe went out also. Gaile Dána found
Cúchulainn and they flung all their twenty-nine spears
together. Not one spear went astray, but Cúchulainn did
the rim-feat with his shield and all the spears sank half
way in the shield. Though they didn't throw wide, therefore, none of their spears was reddened with his blood.
Then Cúchulainn pulled the sword from his Badb's scabbard to cut away the spears and lighten his shield. They
attacked him while he was doing this and aimed their
twenty-nine fists together at his head, and struck him and
bore him down till his face met the ford's sandy gravel.
He uttered his warrior's scream on high, and his cry of
unfair fight, so that every living man in Ulster heard it,
except those that lay asleep. Fiacha leaped from his
chariot when he saw their hands all raised against Cúchulainn and he hacked off all twenty-nine hands.
Cúchulainn said:

'That was help in the nick of time!'

'It wasn't much,' Fiacha said. 'But the compact is
broken now for the Ulster exiles. If a single one of these
gets back to the camp our whole troop of three thousand
will go under the edge of the sword.'

'I swear by the gods,' Cúchulainn said, 'that not one of
them will get there alive while I draw breath.'

Cúchulainn, with the two sons of Ficce — two brave
Ulster warriors who had come to try their strength on the
armies — slew all twenty-nine. So ended this episode of
battle with Cúchulainn on the Táin. On a stone in the
middle of the ford there is the mark where the shield
was thrown and the marks of their fists and knees. Twenty-
nine standing-stones were erected there for them.

XI COMBAT OF FERDIA AND CUCHULAINN

THEY brooded among themselves on the man who might next protect them from Cúchulainn. The four provinces of Ireland talked and argued back and forth about who should go against him at the ford. They all agreed it should be the horn-skinned warrior from Irrus Domnann, the burden unbearable and the rock fatal in the fray, Cúchulainn's own ardent and adored foster-brother. There was not a feat of Cúchulainn's that he lacked, except the *gae bolga*, and they thought he could avoid that and save himself by means of the stuff of horn he had around him. No weapon, no edge, could pierce it.

Medb sent messengers to Ferdia, but he wouldn't come back with them. Then Medb sent poets and bards and satirists to bring the blushes to his cheek with mockery and insult and ridicule, so there would be nowhere in the world for him to lay his head in peace. In dread of being put to shame by these messengers he came back with them to Medb's and Ailill's tent, where it had been pitched on the Táin. Their daughter Finnabair was put beside him. She handed him the goblets and cups, with three kisses for every cup. And at the neck-opening of her shirt she

offered him certain fragrant sweet apples, saying that
Ferdia was her darling and her chosen beloved of the
whole world. When Ferdia was full and in good humour,
Medb said:

'Well now, Ferdia, do you know why you were brought
to our tent?'

'I know the noblest men in Ireland are here,' Ferdia
said. 'Why shouldn't I be here too, as well as these fine
warriors?'

'That isn't the reason,' Medb said. 'but to give you a
chariot worth three times seven bondmaids, with war-
harness enough for a dozen men, and a portion of the
fine Plain of Ai equal to the Plain of Murtheimne. Also
the right to stay forever in Cruachan, with your wine
supplied, and your kith and kin free forever from tax and
tribute. And this leaf-shaped brooch of mine that was
made out of ten score ounces and ten score half-ounces
and ten score cross-measures and ten score quarters of
gold. And Finnabair, my daughter and Ailill's, for your
wife. And my own friendly thighs on top of that if needs
be.'

'No need!' they all cried. 'Those gifts and trophies are
enough.'

'They are certainly very great,' Ferdia said. 'But great
as they are, Medb, I would sooner leave them with you
than go out to fight my own foster-brother.'

'What Cúchulainn said was true, my people,' Medb
said, as though she hadn't heard Ferdia. She knew well
how to stir up strife and dissension.

'What did he say, Medb?' Ferdia said.

'He said he wouldn't count it any great triumph if his
greatest feat of arms were your downfall,' she said.

'He shouldn't have said that. He never knew me slow
or sluggish to fight, night or day. I swear by the gods I'll

be first at the ford of battle tomorrow morning to fight
him.'

'All our blessing! Go and win,' Medb said. 'This is
better than being thought sluggish and slow to arms for
the sake of some loyalty outside our own people. Is it
fitting for him to guard the safety of Ulster because his
mother belonged there, but not for you, the son of a king
of Connacht, to save the province of Connacht?'

So the promise was made, and they chanted together:

> Medb :   'Riches and rings I promise,
>   a share of woods and plain,
>   privilege for your kinsfolk
>   to the end of time.
>   Does it take your breath away,
>   Ferdia mac Damáin?
>   It is yours; accept it
>   — others have.'
>
> Ferdia : 'Give me some surety.
>   I'm no hollow hero,
>   but tomorrow I go to bear
>   a terrible trial.
>   Culann's harsh Hound
>   is not faced lightly.
>   It could be a stern matter,
>   a dire disaster.'
>
> Medb :   'That is no trouble.
>   Pick your own surety
>   from among kings or princes,
>   any hostage you wish
>   — there are men who can ensure
>   whatever you ask.
>   But I know you will kill
>   this man when you meet him.'

Ferdia : 'I'll pick six heroes,
   six and no less,
   before I try my strength
    in front of the armies.
   Grant me this
   and I'll do battle
   with the hard Hound,
    though I'm not his match.'

Medb : 'Take farmers or soldiers,
   or Niaman the slaughterer,
   or choose among the bards
    and you can have them.
   If you so demand
   you may have Morann
   or Coirpre nia Manann
    or our own two sons.'

Ferdia : 'You've a strong tongue, Medb.
   Your kind husband's no curb.
   There's no doubt you are master
    on the mounds of Cruachan!
   By your fame and great force,
   I'll take the speckled silk,
   the gold and the silver
    and all you have promised.
   Give me six princes
   to be my surety
   when I go to my doom
    with the hard Hound.'

Medb : 'Choicest of champions,
   take this round brooch.
   Rest now until Sunday,
    when the fight is due.

> My famous fine warrior,
> all shall be given
> into your hands
>   — the world's greatest jewels,
> and queenly Finnabair,
> the heroes' favourite —
> when the Hound is finished;
>   all yours, Ferdia.'

The great Ulster warrior, Fergus mac Roich, was there while they were bargaining. He went to his tent and said:
'There will be a sad deed done tomorrow morning.'
'What is that?' the people in his tent said.
'The slaying of my good foster-son Cúchulainn.'
'Indeed! who is boasting of that?'
'It is soon told: his own beloved fiery foster-brother, Ferdia mac Damáin. Will one of you, in pity for Cúchulainn, go to him with my blessing,' Fergus said, 'and warn him to flee from the ford tomorrow morning?'
'By my soul,' they said, 'we wouldn't go on such an errand, even though you yourself were due in the ford of battle.'
'Very well,' Fergus said. 'Get the horses, my friend, and yoke the chariot.'
The charioteer rose and got the horses and yoked the chariot. They went out to the ford of battle and found Cúchulainn. Fergus came up and alighted from his chariot.
'You are welcome, friend Fergus,' Cúchulainn said.
'I believe I am,' Fergus said.
'What do you want here?' Cúchulainn said.
'To tell you the warrior who is coming to fight you tomorrow morning,' Fergus said.
'Tell it, then. Let me hear it,' Cúchulainn said.

'Your own foster-brother Ferdia mac Damáin.'

'I swear I don't want this meeting,' Cúchulainn said. 'Not because I fear him but because I love him so much.'

'You would do well to fear him too,' Fergus said. 'He has a skin of horn on him when he fights that no point or blade can pierce.'

'You needn't worry,' Cúchulainn said. 'If he appears at the ford before me, I swear by the vow of my people that his joints and limbs will bend like reeds in the river at the point of my sword.'

So they spoke together there, and chanted:

> Fergus: 'Cúchulainn, you are well met.
> It is time you were astir.
> Ferdia mac Damáin is coming.
> His face is red with rage against you.'

> Cúchulainn: 'Here I stand, an obstacle
> to all the men of Ireland!
> I have stood my ground here
> through countless single combats.'

> Fergus: 'I don't wish to cause unease,
> but — for all your fame, Cúchulainn —
> Ferdia wears a horn-skin
> that no kind of weapon pierces.'

> Cúchulainn: 'When I and skillful Ferdia
> fight it out in the ford together
> we will know before we're done
> whom the fierce blades favour.'

> Fergus: 'His brave arm is strong with rage.
> He holds a blood-reddened sword.
> He is strong as a hundred men
> and safe from every point or edge.'

Cúchulainn : 'Fergus, with your mighty weapons,
              say no more — it is enough.
              There are no odds too much for me
              in any part of Ireland.'

Fergus : 'Cúchulainn, with your crimson sword,
          nothing could please me more
          if you carried off the spoils
          from Ferdia in his pride.'

Cúchulainn : 'I am not a boastful man
              but I swear now — I do not lie —
              I will take the victory
              from Damán mac Dáiri's son.'

Fergus : 'When I turned against Ulster
          it was to avenge a wrong.
          Heroes and war-like men
          left their homes to come with me.'

Cúchulainn : 'Only for Conchobor's pangs
              you'd have found it harder still.
              Medb's journey from Scáil Plain
              would have been a trail of tears.'

Fergus : 'The great trial is at hand,
          the fight with Ferdia mac Damáin,
          with the hard baleful bitter spear,
          Hound of Culann, for your share.'

Cúchulainn said :
'Is this the reason you came, Fergus, my friend?'
'It is,' Fergus said.
'It is as well,' Cúchulainn said, 'it was no one else of
the men of Ireland came to warn me about a single

warrior. All the four provinces of Ireland would have been needed to save him!'

Then Fergus went off to his tent.

'What will you do tonight,' Laeg said to Cúchulainn.

'What do you mean?' Cúchulainn said.

'When Ferdia comes to attack you, he will be washed and bathed, with hair nicely plaited and freshly trimmed, and the four provinces of Ireland will come with him to watch the fight. I think you should go where you will get the same attention, where sweet-haired Emer is waiting in Cairthenn Cluana-Dá-Dam, the Meadow of the Two Oxen, at Sliab Fuait.'

Cúchulainn went there that night and stayed with his wife. No more is said here about that.

As to Ferdia, he went to his tent and told his followers of the pledge he had given Medb, to wage single combat on Cúchulainn the next day or fight six warriors failing that. He told them of the equal pledge he had got from Medb, to have the same six warriors sent to make sure she fulfilled her promises if he killed Cúchulainn. The people in Ferdia's tent that night were gloomy and oppressed. They felt certain that if these two shafts of battle of the world met together there would be a double downfall: it might even happen that their own lord would fall. For it was no easy thing facing Cúchulainn on the Táin.

Great anxieties weighed on Ferdia's spirit that night and wouldn't let him sleep. Among these was the thought that he might lose the treasures and the girl, in combat with this one man, while if he didn't fight that man next day he would have to fight the six champions. But a greater worry than all was the knowledge that his life and his head would never again be in his own hands if he once appeared at the ford before Cúchulainn.

Ferdia got up early next day.

'Now, my friend,' he said, 'bring our horses and yoke the chariot.'

'I swear,' the charioteer said, 'we'd do better not to take this journey.'

Ferdia talked with his charioteer and encouraged him, and they chanted:

> Ferdia :   'Let us go to do battle
> with the man waiting
> down at the ford
>   where the Badb will screech.
> Let us meet Cúchulainn.
> I'll pierce his slight body
> and pass the spear through him
>   and bring him death.'

> Charioteer :   'These are cruel threats.
> Better stay here
> or one will die
>   — a sudden parting.
> You are going to disaster
> before all Ulster.
> It will long be remembered.
>   Beware, if you go.'

> Ferdia :   'You waste your breath.
> Is it warrior's work
> to be shy or meek?
>   I won't hold back.
> Silence, my friend,
> have courage to the end.
> Not fearful but firm
>   let us go to do battle.'

The charioteer got the horses and yoked the chariot and they left the camp.

'Wait a moment,' Ferdia said. 'It isn't right to leave without bidding farewell to the men of Ireland. Turn the horses and chariot round to face them.'

The charioteer turned the horses and chariot round three times and faced the men of Ireland. He passed close to Medb as she was making water on the floor of the tent.

'Are you asleep still, Ailill?' Medb was saying.

'No,' Ailill said.

'Do you hear your new son-in-law bidding you farewell?'

'Is that what he's doing?' Ailill said.

'It is,' Medb said, 'but I swear by the vow of my people that the man making his farewell there won't be coming back to us on his own feet.'

'We have done well with the marriage agreement, anyhow,' Ailill said, 'if he kills Cúchulainn. It is all the same to us if they both die. Still, it might be better if Ferdia escaped.'

Ferdia proceeded to the ford of battle.

'See is Cúchulainn at the ford,' Ferdia said.

'He isn't there,' the charioteer said.

'Look well,' Ferdia said.

'Cúchulainn is not such a little speck that you couldn't see him if he was there,' the charioteer said.

'True enough,' Ferdia said. 'But Cúchulainn never had a real warrior, a proper man, come against him on the Táin Bó Cuailnge until today. As soon as he heard us coming he vanished from the ford.'

'It is a great shame to slander Cúchulainn·in his absence,' the charioteer said. 'Do you not remember when you were fighting the harsh and grizzled Germán Garblas above the borders of the Tyrrhene Sea, and you left your sword

with the enemy army? It was Cúchulainn who killed a
hundred warriors to reach it and bring it back to you.
Do you remember where we stayed that night?'

'No,' Ferdia said.

'At the house of Scáthach's steward,' the charioteer
said, 'and you went before us into the house first, full of
pride and haughtiness. The monstrous steward gave you
a blow of his three-pronged flesh-fork in the small of the
back, and sent you flying out like a stone past the door.
Cúchulainn went in and struck the brute a blow of his
sword and cut him in two. Then I was your steward while
you stayed there. If we could bring that day back, you
wouldn't say you were a better warrior than Cúchulainn.'

'You did wrong, my friend,' Ferdia said, 'not to remind
me of this before. I wouldn't have come looking for this
fight. Pull the shafts of the chariot along beside me now,
and put the skin covering under my head and let me
sleep for a while.'

'Alas for your rest here,' the charioteer said, 'you
would sleep as well in the path of a stag hunt!'

'Why, boy? Can you not keep watch for me?'

'I can,' the charioteer said. 'I'll see them and give
warning before they arrive, from east or west — unless
they come at you out of the clouds and mists.'

The chariot-shafts were pulled along by his side, and
the skin covering put under his head, but he couldn't
sleep, even a little.

Cúchulainn was saying meanwhile:

'Well, friend Laeg, bring the horses, yoke the chariot.
If Ferdia is waiting he must be wondering what keeps us.'

The charioteer got up and brought the horses and
yoked the chariot. Cúchulainn got into the chariot and
they pressed on toward the ford. Ferdia's charioteer wasn't
watching long when he heard the creaking of the chariot

as it drew near. He woke his master and made this chant:

'I hear a chariot creaking.
I see its yoke of silver
and the great trunk of a man
    above the hard prow.
The shafts jut forward,
they are approaching us
by the place of the tree-stump,
    triumphant and proud.

There's a skilled Hound at the helm,
a fine chariot-warrior,
a wild hawk hurrying
    his horses southward.
Surely it is Cúchulainn's
chariot-horses coming.
Who says he is not
    coming to our defeat?

I had a dream last year:
whoever, at the time appointed,
opposes the Hound on the slope,
    let him beware.
The Hound of Emain Macha,
in all his different shapes,
the Hound of plunder and battle
    — I hear him, and he hears.'

'How does Cúchulainn look?' Ferdia said to his charioteer.

'He and his charioteer look as if all the men of Ireland
meant nothing to them,' he said.

'Enough, my friend,' Ferdia said, 'you praise him too

much. Get my weapons ready to meet him at the ford.'

'I feel if I turned my head his chariot-shafts would stick in my neck!'

'Why do you praise Cúchulainn so much?' Ferdia said. 'He hasn't paid you anything.'

Then he chanted:

> 'It's your help I need now,
> not this false friendship.
> Enough of your praises.
>   We're all the same dead!
> Let Cuailnge's great warrior
> come in his glory.
> We'll dispose of him
>   and cut him down.'

Charioteer: 'When Cuailnge's great warrior
> travels in his glory
> it will be here, toward us,
>   not away in flight.
> Give praise where it is due.
> Shrewdly, not slyly,
> he is hurrying toward you
>   — a thunderclap!'

Ferdia: 'There'll be blows between us!
> You haven't stopped praising him
> since we set out.
>   Why do you favour him?
> Even as they challenge him
> everyone praises him.
> But a sorry lot so far
>   has gone to meet him.'

They met in the middle of the ford not long after that and Ferdia said to Cúchulainn:

'You are welcome, Cúchulainn.'

'I could trust your welcome once,' Cúchulainn said, 'but I don't trust it now. Anyway,' he said, 'it is for me and not you, Ferdia, to bid welcome: this is my home-land, you are the intruder. And you are wrong to challenge me to combat. It would suit me better to challenge you: you have driven out our women and young men and boys, and our troops of horses, our herds and our flocks and all our goods.'

'Enough, Cúchulainn,' Ferdia said. 'What brings you at all to meet me in this war-like combat? When we were with Scáthach and Uathach and Aife, you were only my body-servant, who fixed my spears and made my bed.'

'That is true,' Cúchulainn said, 'but I did it then because I was young and small. You can't call me that now. There isn't a warrior in the world now that I'm not able for.'

Then each bitterly reproached the other and they broke off their friendship and Ferdia chanted, with Cúchulainn answering:

> 'What brings you here, Squinter,
> to try my strength!
> Through the steam of your horses
>   I'll reach and redden you.
> You'll regret you came.
> You're a fire without fuel.
> You'll need plenty of help
>   if you ever see home.'

Cúchulainn: 'Like a great boar
> before his herd,
> I'll overwhelm you
>   before these armies.
> I'll push you and punish you
> to the last of your skill,

                    and then bring down
                        havoc on your head!'

Ferdia :   'It is I who will kill,
            I who will destroy,
            I who will drive
                Ulster's hero to flight
            before all eyes.
            By my doing
            they'll rue their loss
                early and late.'

Cúchulainn :   'Must we start our fight
            groaning over corpses?
            Come what may
                let us enter the ford
            to meet death before the hosts
            with bloody spear-blade
            or the savage sword
                if our time is come.'

Ferdia :   'Attack then, if we must.
            Before sunset and nightfall
            I'll fight you at Bairche
                in bloody battle.
            Men of Ulster will cry out:
            "Death has seized you!"
            The terrible sight
                will pierce them through.'

Cúchulainn :   'You have reached your doom,
            your hour is come.
            My sword will slash
                and not softly.

When we meet you will fall
at a hero's hands.
Never again
    will you lead men.'

Ferdia : 'Little bush, you have boasted
and threatened enough.
You'll find no mercy
    or victory here.
I know you well —
a clumsy and feeble
chicken-hearted
    trembling boy.'

Cúchulainn : 'While we stayed with Scáthach
we went as one,
with a common courage,
    into the fight.
My bosom friend
and heart's blood,
dear above all,
    I am going to miss you.'

Ferdia : 'You make much of yourself,
but the fight is to come.
I'll have spiked your head
    when the cock crows.
Cúchulainn of Cuailnge
has lost his wits
and will suffer for it.
    The guilt is yours.'

'Ferdia, you did wrong to come fighting with me,'
Cúchulainn said. 'It is only Ailill and Medb's ill-doing and

meddling. It has gone badly with all who came against me — I killed them all. And it can do you no good. You too will fall.'

He spoke further and Ferdia listened :

'Ferdia son of Damán,
noble warrior, do not come.
You will suffer more than me
and bring sorrow to your company.

Do not come — and in the wrong —
or here you'll find your resting-place.
How can it be that you alone
could escape my fatal rage?

I'll overwhelm you with my feats
—- despite your horn-skin and red rage.
Son of Damán, you'll never have
the girl that you are boasting of.

Medb's daughter Finnabair,
for all the fairness of her form
and all the sweetness of her shape,
will never yield to your assault.

Finnabair, the royal daughter
— she is nothing but a snare.
She played false with the others
and ruined them as she ruins you.

Don't break our friendship and our bond,
don't break the oath we made once,
don't break our promise and our pledge.
Noble warrior, do not come.

This is the same girl who was promised
falsely to fifty men.
They got nothing but my spear
as I showed them to their graves.

Ferbaeth, they said, was brave enough
and had a houseful of fine heroes,
but a short moment quenched his fire,
I finished him with one throw.

Srúbdaire found a bitter end.
A hundred women held him dear.
There was a time his fame was high,
but neither wealth nor weapons saved him.

If they had offered her to me,
if I were the one that Medb smiled at,
I wouldn't think to do you harm
or touch the least part of your flesh.

'That is the reason you shouldn't come to fight me,
Ferdia,' Cúchulainn said. 'When we were with Scáthach
and Uathach and Aife we always set out together to the
battles and battlefields, to the strife and the struggle and
the forests and deserts and dark mysteries.'

He spoke further:

'Fast friends, forest-companions,
we made one bed and slept one sleep
in foreign lands after the fray.
Scáthach's pupils, two together,
we'd set forth to comb the forest.'

Ferdia said:

'Cúchulainn, you bear your cunning lightly,
but I have mastered the same trade.

Our friendship is finished, through foul play.
Prepare to face your first defeat.
Forget that we were foster-brothers.
Squinter, you are past help!

'We have talked too much,' Ferdia said. 'What weapons will we use today, Cúchulainn?'

'You have the choice of weapons until nightfall,' Cúchulainn said. 'You reached the ford first.'

'Do you remember,' Ferdia said, 'the very last feats we learned under Scáthach and Uathach and Aife?'

'I remember them well,' Cúchulainn said.

'Let us set to, since you remember.'

They set to with those last feats. They took up their two finely-marked feat-playing shields and their eight shields with the sharp rims, their eight darts and their ivory-hilted straight swords and their eight small ivory darts that flew between them like bees on a pleasant day. They threw nothing that didn't hit. And they were busy with these feats from the grey of early morning to the middle of the day, attacking each other and bringing each other's many feats to nothing with the knobs and bosses of their feat-playing shields. But no matter how finely they threw they fended-off just as finely, so that neither drew blood from the other during all that time.

'Let us break off now with these weapons, Cúchulainn,' Ferdia said. 'We'll settle nothing this way.'

'Very well, let us break off if it is time,' Cúchulainn said.

They broke off. They threw their feat-playing gear into their charioteers' arms.

'What weapons will we use next, Cúchulainn?' Ferdia said.

'You still have the choice of weapons until nightfall,'

Cúchulainn said, 'since you were first at the ford.'

'Then,' Ferdia said, 'let us try our strong, smooth-polished slender spears, bound with the tight flax.'

'Very well, let us try them,' Cúchulainn said.

They took up their two tough shields, well matched for strength, and their strong, smooth-polished slender spears tightly bound with flax. They hurled their spears at each other from the middle of the day until the evening sunset. And finely though they fended-off, they cast more finely still, and wounded and gored and bloodied each other for that length of time.

'Cúchulainn, let us break off now from this,' Ferdia said.

'Very well,' Cúchulainn said, 'let us break off if it is time.'

They broke off and flung their weapons into their charioteers' arms. They came up to each other and each put his arm round the other's neck and gave him three kisses. Their horses passed that night in the same paddock and their charioteers by the same fire. Their charioteers made up fresh beds of rushes for them, with rests for their heads, as is right for wounded men. Men of healing and medicine came to heal them and make them whole and dropped wholesome, healing plants and herbs into their stabs and cuts and gashes and countless wounds. As many wholesome, healing plants and herbs as were put on Cúchulainn's stabs and cuts and gashes and countless wounds, he sent the same over to Ferdia on the westward side of the ford, so that the men of Ireland couldn't say, if he killed Ferdia, that he had won because he got more care. Ferdia, out of all the food and the health-giving, stimulating, delicious drinks that the men of Ireland gave him, sent an equal share over to Cúchulainn on the north-ward side of the ford, for there were more supplying

Ferdia with food than Cúchulainn. All the men of Ireland were supplying Ferdia, because he was protecting them from Cúchulainn, while only the people of Breg Plain were supplying Cúchulainn. Each day, when night fell, they used to come and talk to him.

They stayed so that night and got up early next day and came out to the ford of battle.

'What weapons will we use today, Ferdia?' Cúchulainn said.

'You have the choice of weapons until nightfall,' Ferdia said. 'I had my choice yesterday.'

'Then,' Cúchulainn said, 'let us try our big burdensome stabbing-spears. We may bring the end nearer today with our stabbing than with yesterday's spear-throwing. Let our horses be brought and the chariots yoked. Today we'll fight with horse and chariot.'

'Let us begin,' Ferdia said.

So that day they took up their two solid broadshields and their big burdensome stabbing-spears and began piercing and drilling each other and felling and overwhelming from the grey of early morning until the evening sunset. If ever birds in flight could pass through men's bodies they could have passed through those bodies that day and brought bits of blood and meat with them out into the thickening air through the wounds and gashes. When the sun set that evening the horses were spent and the charioteers dazed and the high heroes themselves were at an end.

'Let us break off now from this, Ferdia,' Cúchulainn said, 'for our horses are spent and our charioteers are dazed, and if they are finished why shouldn't we be finished too?'

He said further :

'Why suffer the chariots' plunging
or struggle like Fomorian giants?
Hobble the horses.
Let the turmoil die away.'

'Very well, let us break off if it is time,' Ferdia said.
They broke off and threw their weapons into their
charioteers' arms. They came up to each other and each
put his arm round the other's neck and gave him three
kisses. Their horses passed that night in the same paddock
and their charioteers by the same fire, and their charioteers
made fresh beds of rushes for them, with rests for their
heads, as is right for wounded men. Men of healing and
medicine came to watch over and guard them and mind
them that night. So hideous were their stabs and cuts and
gashes and countless wounds that nothing could be done
but lay magic amulets on them and say spells and incanta-
tions to stop the spurts and spouts of blood. For every
amulet or spell or charm that was laid on Cúchulainn's
cuts and gashes he had the same sent to Ferdia across on
the westward side of the ford, and Ferdia, out of all the
food and health-giving, stimulating, delicious drinks that
the men of Ireland gave him, sent an equal share over to
Cúchulainn on the northward side of the ford.

They stayed so that night and got up early next day,
and came out to the ford of battle. That day Cúchulainn
saw an aspect of evil and a dire darkness over Ferdia.

'You have a dreadful look today, Ferdia,' he said. 'A
shadow has fallen on your hair overnight and your eye
has grown dull. All your fine shape and strength and
structure are gone.'

'It is not for any terror or dread of you,' Ferdia said.
'There isn't a warrior in Ireland that I can't beat off.'

Cúchulainn lamented and sorrowed. He made this

chant, with Ferdia answering:

> 'Ferdia, is it you I see?
> Now I know it was your doom
> when a woman sent you here
> to fight against your foster-brother.'

Ferdia: 'Cúchulainn, you are wise enough,
a true hero, a true warrior.
You know that everyone must come
to the sod that is his last bed.'

Cúchulainn: 'Medb's daughter Finnabair
— whatever beauty she may have —
was never promised you for love,
but so that you would use your strength.'

Ferdia: 'My strength has been well used by now,
Hound of the sweet discipline.
Never to this very day
did I find braver, or hear of one.'

Cúchulainn: 'Yours is the blame for what must come,
son of Damán mac Dáiri
— coming, at a woman's word,
to cross swords with your foster-brother.'

Ferdia: 'Sweet Hound, if we part now
— though foster-brothers—without a fight,
think of my ill-fame and shame
at Cruachan before Ailill and Medb.'

Cúchulainn: 'There is no man that ever ate,
no man that was ever born,
no joyous son of king or queen,
for whose sake I would do you harm.'

Ferdia :   'Cúchulainn, tide of bravery,
           I know that Medb has ruined us.
           You will win victory and renown
           and no one think you were at fault.'

Cúchulainn : 'My high heart is a knot of blood,
              my soul is tearing from my body,
              I'd rather face a thousand fights,
              Ferdia, than this fight with you.'

'You may blame me all you like today,' Ferdia said.
'What weapons will we use?'

'You have the choice of weapons today until nightfall,'
Cúchulainn said. 'I chose yesterday.'

'Then,' Ferdia said, 'let us take up our massive stroke-dealing swords. We may bring the end nearer today with our hacking than with yesterday's stabbing.'

'Let us begin, then,' Cúchulainn said.

So that day they took up their two great full-length shields and their massive stroke-dealing swords and began hacking and hewing and striking and destroying, and cutting bits and pieces the size of baby's heads from each other's shoulders and backs and flanks.

They hacked at each other in this way from the grey of early morning until the evening sunset.

'Cúchulainn, let us break off from this,' Ferdia said.

'Very well,' Cúchulainn said.

They broke off and flung their weapons into their charioteers' arms.

They had met that day, two solid and satisfied men, lively and serene. But they parted that night woeful and weary, two wasted men. And it wasn't in the same paddock that their horses passed that night, nor at the same fire their charioteers.

They stayed so that night. Ferdia got up early next day and came out alone to the ford of battle, for he knew that this day would decide the fierce struggle, and that one of them, or both, would fall. He put on his war-like battle-harness before Cúchulainn came out to meet him. This was his battle-harness: a filmy girdle of silk with a speckled-gold hem next his bright skin, a dark supple apron of leather over that on the outside, and a stout strong stone outside that again, the size of a millstone. That day, for fear and dread of the *gae bolga*, he put a deep and sturdy apron of twice-smelted iron over the stout strong stone like a millstone. He set on his head his war-like crested battle-helmet, finely decorated with forty precious carbuncles and inlaid with red enamel and crystal and carbuncle and gleaming stones from the East. He took in his right hand his furious spear, stout and fierce. In his left he took his battle-sword with its gold grip and its hilt of red-gold. On the curve of his back he took his handsome huge shield with the great red-gold knob in the middle and another fifty knobs around it, each big enough to hide a prize boar. That day Ferdia did a thousand thrilling feats on high, multiple and miraculous, that no-one had ever taught him — not his foster-mother or foster-father, nor Scáthach nor Uathach nor Aife — but drawn from him that day at the thought of Cúchulainn.

Cúchulainn came to the ford then and saw Ferdia's thousand thrilling feats on high, multiple and miraculous.

'Look, friend Laeg, at the thousand feats Ferdia does on high, multiple, miraculous — thrilling! He is going to use them all on me today. If my defeat seems near at any time, you must abuse and insult and mock me to make my anger rise. But if ever his defeat seems near tell me that, and praise and encourage me to raise my spirits.'

'I will, Cúchulainn,' Laeg said.

Then Cúchulainn too put on his war-like battle-harness and did a thousand thrilling, multiple and miraculous feats on high that he also had learned from no-one — not Scáthach nor Uathach nor Aife. Ferdia saw those feats and knew they were all for him.

'What weapons shall we use, Ferdia?' Cúchulainn said.

'You have the choice of weapons until nightfall,' Ferdia said.

'Very well, let us try fighting in ford water,' Cúchulainn said.

'Let us try that,' Ferdia said.

Ferdia, though he spoke lightly, knew that it was the worst thing for him, because Cúchulainn destroyed every hero and high warrior that ever fought him in ford water. Still, marvellous deeds were done by the two heroes that day in the ford — those two first and foremost, those two chief chariot-warriors of the west, those two blazing torches of bravery in Ireland, those two lavish and liberal gift-scatterers of the whole northwest of the world, those two keys to Ireland's valour, flung together from afar by the ill-doing and meddling of Ailill and Medb. They began working their feats on each other there from the grey of the dewy dawn until high noon. At noon the men's madness mounted and they drew closer to each other.

Cúchulainn sprang straight from the brink of the ford on to the shield-knob of Ferdia mac Damáin to strike down over the edge of the shield at his head. Ferdia struck the shield a blow of his left elbow that sent Cúchulainn away back from him like a bird past the brink of the ford. Cúchulainn sprang again from the brink onto the knob of Ferdia's shield, to strike down at his head over the edge of the shield. But Ferdia struck the shield a blow of his left knee that sent Cúchulainn away back from him like a little boy past the brink of the ford.

Laeg saw this.

'Well, now!' he said. 'Your enemy shook you then as easily as a loving mother slaps her son! He tossed you aside as if he was rinsing a cup in a tub! He crushed you like a mill crushing fine malt! He went through you like a drill through an oak! He bound you in knots like a creeper entangling a tree! He pounced on you like a hawk on a little bird! From this day onward, my devilish little half-sprite,' Laeg said, 'you have no right or claim or title to great deeds or daring.'

At that Cúchulainn rose up for the third time, quick as the wind, swift as a swallow, in a storm of strength and dragonish fury, and landed on the knob of Ferdia's shield and tried to strike down at him over the shield-rim. But that battle-warrior gave a shake of his shield that sent Cúchulainn off, as though he had never landed on it, into the middle of the ford. Cúchulainn warped in his fury-spasm; he blew up and swelled like a bladder full of breath and bent himself in a fearful hideous arch, mottled and terrifying, and the huge high hero loomed straight up over Ferdia, vast as a Fomorian giant or a man from the sea-kingdom.

Then they fought together so closely that their heads touched at the top and their feet at the bottom and their hands in the middle around the edges and knobs of their shields. So closely they fought that their shields split and burst from rim to belly: so closely they fought that their spears bent and collapsed, worn-out from the tips to the rivets: so closely they fought that their shield-rims and sword-hilts and spear-shafts screamed like demons and devils and goblins of the glen and fiends of the air: so closely they fought that they drove the river off its course and out of its bed, leaving a dry space in the middle of the ford big enough for the last royal burial-ground of a king or queen — not a drop of water on it except what the two heroes and high warriors splashed there in their trampling and slithering in the ford: so closely they fought that the horses of the men of Ireland broke loose in panic and terror, rearing and raving, and broke their shackle-hoops and hobbles and reins and ropes, so that the women and children, the infants, the ill and the imbeciles broke out southwestward from the camp of the men of Ireland.

Then, while they were busy with the sharp sword-edges, Ferdia got a single fatal chance at Cúchulainn, and dealt him a stroke of his ivory-hilted straight-sword and buried it in his breast. The blood gushed over his belt and the ford grew crimson with the battle-warrior's body-gore. Cúchulainn could bear it no longer — all Ferdia's ruinous strokes of strength, his strokes downward and across. And he called out to Laeg mac Riangabra for the *gae bolga*. Ferdia heard Cúchulainn calling for the *gae bolga*, and he dropped his shield to cover his lower body. Then Cúchulainn took his short javelin and hurled it from the middle of his palm over the rim of Ferdia's shield and the edge of his horn-skin, driving it through him so that it pierced the heart in his breast and showed half its length out through his back. Ferdia raised up the shield to cover his upper body, but it was too late. The charioteer sent the *gae bolga* down the stream.

'Beware the *gae bolga*,' he said.

Cúchulainn caught it in the fork of his foot and sent it casting toward Ferdia and it went through the deep and sturdy apron of twice-smelted iron, and shattered in three parts the stout strong stone the size of a mill-stone, and went coursing through the highways and byways of his body so that every single joint filled with barbs.

'That is enough now,' Ferdia said, 'I'll die of that. There is strength in the thrust of your right foot. It is wrong I should fall at your hand.'

He said:

> 'Hound of the bright deeds,
> you have killed me unfairly.
> Your guilt clings to me
> as my blood sticks to you.
>
> By the way of deceit
> no good can come.
> I am struck dumb.
> I am leaving this life.
>
> My ribs are crushed in,
> my heart is all blood.
> I have not fought well.
> Hound, I am fallen.'

Cúchulainn ran toward him and clasped his two arms round him and carried him — weapons, armour and harness — north across the ford with him so that the spoils would be to the north of the ford and not westward of it where the men of Ireland were. Cúchulainn set Ferdia down on the ground and there, by Ferdia's head, fainted away in a cloudy trance. Laeg saw this, and how all the men of Ireland rose to attack him.

'Get up, now, Cúcuc!' Laeg said. 'The men of Ireland are coming to attack us, and they are not thinking of single combat now that you have killed Ferdia mac Damáin meic Dáiri.'

'My friend, why should I rise,' he said, 'and this one fallen by my hand?'

The charioteer spoke to him, with Cúchulainn answering:

> 'Rise up, slaughter-hound of Emain!
> You must recover, have more spirit.
> You have felled Ferdia of the hosts
> — a dire combat, god of doom!'

Cúchulainn : 'What have I to do with spirit?
> Stupor and sorrow weigh me down
> after the deed that I have done,
> this corpse that I have hacked so harshly.'

Laeg : 'You have nothing to regret;
> indeed you ought to boast of it.
> He has stained his spear in you
> and left you streaming, and near death.'

Cúchulainn : 'What matter? He could have taken off
> my leg, or my very arm.
> Alas, Ferdia of the steeds
> will never draw another breath.'

Laeg : 'The women of the Craebruad
> wouldn't have it otherwise.
> Ferdia dead and you alive
> — that separation they can bear.'

Cúchulainn : 'From the first day I left Cuailnge
> to come against the mighty Medb
> she has had carnage and renown,
> with all the warriors I've slain.'

Laeg : 'You have had no sound sleep
> since you stopped the great Táin.
> Because there were so few to help
> you woke early many a morning.'

Cúchulainn began mourning and lamenting Ferdia there, and said:

'Alas Ferdia! Woe for you that you didn't listen, before we fought together, to somebody who knew my high, brave deeds; woe for you that Laeg mac Riangabra didn't chide you with memories of our fostering together; woe for you that you rejected Fergus's well-meant warning; woe for you that proud, kind Conall, much honoured in arms, didn't help with word of our fostering together. Those are men who wouldn't run to you with news of the wants and wishes or the false promises of any fair-headed Connacht woman; those are men who knew that none of human birth, until the day of doom, can ever match the heavy, high deeds that I do against Connacht with shield or shield-rim, sword or dart, draughts or chess, horse or chariot. Never will hand of warrior hack the flesh of a hero like the honoured heir Ferdia. Never will the red-mouthed Badb screech like this at the shield-bright sheltering hosts in the gap of battle. Never till the day of doom will any one fighting for Cruachan get the bargain you got, crimson-visaged son of Damán,' Cúchulainn said.

Cúchulainn got up from beside Ferdia's head.

'Well, Ferdia,' Cúchulainn said, 'it was a great doom and desolation that the men of Ireland wished on you when they sent you to do battle and combat with me. It is no light thing to struggle and strive with Cúchulainn on the Táin Bó Cuailnge.'

He said:

> 'Ferdia, dead by their deceit,
> our last meeting I lament.
> You are dead and I must live
> to mourn my everlasting loss.

When we were away with Scáthach
learning victory overseas,
it seemed our friendship would remain
unbroken till the day of doom.

I loved the noble way you blushed,
and loved your fine, perfect form.
I loved your blue clear eye,
your way of speech, your skillfulness.

Your like, Damán's crimson son,
never moved to the tearing fray,
never was seized with manly wrath
nor bore shield on his broad back.

Never till this very day,
Ferdia, did I ever find
your match for great deeds in battle
since I slew Aife's one son.

Medb's daughter Finnabair,
whatever beauty she may have,
was an empty offering,
a string to hold the sand, Ferdia.'

Cúchulainn stayed staring there at Ferdia.
'Well, friend Laeg,' Cúchulainn said, 'strip Ferdia now.
Take off his gear and garments. Let me see the brooch he
fought this furious battle for.'
Laeg came and stripped Ferdia and took off his gear
and garments, and showed him the brooch. Cúchulainn
mourned and lamented:

'Ferdia of the hosts
and the hard blows, beloved
golden brooch, I mourn
your conquering arm

and our fostering together,
a sight to please a prince;
your gold-rimmed shield,
your slender sword,

the ring of bright silver
on your fine hand,
your skill at chess,
your flushed, sweet cheek,

your curled yellow hair
like a great lovely jewel,
the soft leaf-shaped belt
that you wore at your waist.

You have fallen to the Hound,
I cry for it, little calf.
The shield didn't save you
that you brought to the fray.

Shameful was our struggle,
the uproar and grief!
O fair, fine hero
who shattered armies
and crushed them under foot,
golden brooch, I mourn.'

'Now, friend Laeg,' Cúchulainn said, 'cut Ferdia open
and take the *gae bolga* out of him. I must have my
weapon.'

Laeg came up and cut Ferdia open and took out the
gae bolga. Cúchulainn saw his weapon crimson and bloody
from Ferdia's body and said:

'Ill-met, Ferdia, like this
— you crimson and pale in my sight
and stretched in a bed of blood,
I with my weapon unwiped.

When we were beyond the sea,
Scáthach's and Uathach's pupils,
who thought of such pale lips
or weapon-struggle between us?

I remember when Scáthach lifted
her sharp harsh cry:
"Germán Garbglas is coming!
Forward to the furious fray!"

Then I said to Ferdia
and to Lugaid of the lavish hand
and to fond, foolish Ferbaeth:
"Let us go to meet Germán."

At the battle-rock on the slope
above the Lake of Envy
we took out four hundred men
from the Islands of Victory.

I stood with fierce Ferdia
in the door of Germán's fort.
I killed Rinn mac Niuil,
he Ruad mac Forniuil.

Ferbaeth killed Bláth mac Colbaí
of the red sword, on the slope.
Grim, swift Lugaid slew
Mugairne from the Tyrrhene Sea.

We went in and I slew there
four times fifty raging men.
Ferdia killed Dam Dreimend
and Dam Dílenn — a cruel crew.

We levelled Germán's cunning fort
above the wide, glittering sea
and took Germán himself alive
to Scáthach of the great shield.

Our famous foster-mother bound us
in a blood pact of friendship,
so that rage would never rise
between friends in fair Elga.

Sad and pitiful the day
that saw Ferdia's strength spent
and brought the downfall of a friend.
I poured him a drink of red blood!

If you had met your death then
fighting with Greek warriors,
I wouldn't have outlasted you,
I would have died at your side.

Misery has befallen us,
Scáthach's two foster-sons
— I, broken and blood red,
your chariot standing empty.

Misery has befallen us,
Scáthach's two foster-sons
— I, broken and blood-raw,
you lying stark dead.

Misery has befallen us,
Scáthach's two foster-sons
— you dead and I alive.
Bravery is battle-madness!'

'Well, Cúcuc,' Laeg said, 'let us leave the ford now. We
have been here too long.'

'Very well, let us leave it, friend Laeg,' Cúchulainn
said. 'All the struggles and contests that I ever fought
seem only playful games now after my struggle with
Ferdia.'

And he said these words:

'It was all play, all sport,
until Ferdia came to the ford.
A like learning we both had,
the same rights, the same belongings,
the same good foster-mother
— her whose name is most honoured.

All play, all sport,
until Ferdia came to the ford.
The same force and fury we had,
the same feats of war also.
Scáthach awarded two shields,
one to me, one to Ferdia.

All play, all sport,
until Ferdia came to the ford.

Misery! A pillar of gold
I have levelled in the ford,
the bull of the tribe-herd,
braver than any man.

All play, all sport,
until Ferdia came to the ford
— fiery and ferocious lion,
fatal, furious flood-wave!

All play, all sport,
until Ferdia came to the ford.
I thought beloved Ferdia
would live forever after me
— yesterday, a mountain-side;
today, nothing but a shade.

I have slaughtered, on this Táin,
three countless multitudes:
choice cattle, choice men,
and horses, fallen everywhere!

The army, a huge multitude,
that came from cruel Cruachan
has lost between a half and a third,
slaughtered in my savage sport.

Never came to the battle-field,
nor did Banba's belly bear,
nor over sea or land came
a king's son of fairer fame.'

XII   ULSTER RISES FROM ITS PANGS

THE ARMIES went off southward from Ferdia's Ford. Cúchulainn lay there sick. Senoll Uathach, the Hideous, and the two sons of Ficce were the first to reach him. They bore him back with them to Conaille, where they nursed his wounds and bathed them in the waters of the river Sas, for ease, the river Búan for steadfastness, Bithslán for lasting health, the clear Finnglas, the bright Gleóir, the dashing Bedc; in Tadc, Talamed, Rinn and Bir, in the sour Brenide and narrow Cumang; in Celenn and Gaenemain, Dichu, Muach and Miliuc, Den, Deilt and Dubglas.

While Cúchulainn was washing in those waters, the armies continued and pitched camp at Imorach Smiromrach, close by the Mash of Marrow, of which you shall hear. Mac Roth left the armies and went northward to keep watch on the men of Ulster. He went as far as

Sliab Fuait to see if any were following. He brought back news that he saw only one chariot.

'I saw a chariot crossing the plain from the north,' Mac Roth said. 'The man had silvery-grey hair and carried no weapon but a silver spike in one hand. His chariot was coloured bright as the May. He was goading the chariot-eer as well as the horses, as though he felt he would never catch up with the armies. A brindled hunting dog ran in front of him.'

'Who would that be, Fergus?' Ailill said. 'Do you think it was Conchobor or Celtchar?'

'No' Fergus said. 'I believe it is Cethern, Fintan's son, a man of generosity and a bloody blade.'

Fergus was right. Cethern hurled himself directly at the camp and slaughtered many men. But he himself was wounded badly. He came back from the battle toward Cúchulainn with his guts around his feet, and Cúchulainn pitied his wounds.

'Get me a healer,' Cethern said to Cúchulainn.

A bed of fresh rushes was fixed for him, with a pillow, and Cúchulainn sent Laeg to the enemy camp, to ask Fiacha mac Fir Febe for a healer, and to say he would kill them all if they didn't come to look at Cethern — no matter where they hid themselves, even under the earth. The healers grew worried at this, for there was no one in the camp that he hadn't hit; but they went out to see him. The first healer came up and examined him.

'You won't survive this,' he said.

'Then neither will you!' Cethern cried, and struck him with his fist, and his brains splashed over his ears. He killed fifty healers, some say, in the same way, though others say he killed only fifteen. The last of them got a glancing blow and fell stunned. Cúchulainn saved his life.

Cúchulainn said to Cethern:

'You had no right to kill those healers. We'll get no one to come to you now.'

'They had no right to give me bad news.'

They sent for the holy healer Fingin, Conchobor's own healer, to examine Cúchulainn and Cethern. Fingin was well aware of the great sufferings of Cúchulainn and Cethern, and soon they saw his chariot coming. Cúchulainn went up to him and said:

'Watch out for Cethern.' (Indeed it would have been foolish not to, when he had already killed fifteen other healers).

Fingin went up and studied him from a distance.

'Examine me,' Cethern said. 'This great wound here looks grave. What made it?'

'A vain, arrogant woman gave you that wound,' Fingin said.

'I believe you are right,' Cethern said. 'A tall, fair, long-faced woman with soft features came at me. She had a head of yellow hair, and two gold birds on her shoulders. She wore a purple cloak folded about her, with five hands' breadth of gold on her back. She carried a light, stinging, sharp-edged lance in her hand, and she held an iron sword with a woman's grip over her head — a massive figure. It was she who came against me first.'

'Then I'm sorry for you,' Cúchulainn said. 'That was Medb of Cruachan.'

'This next,' the healer said, 'was a light, half-hearted wound from some kinsman. It won't kill you.'

'That is true,' Cethern said. 'A warrior with a curved scallop-edged shield came at me. He had a curve-bladed spear in his hand and an ivory-hilted, iron-bladed sword in three sections. He wore a brown cloak wrapped around him, held with a silver brooch. He took a slight wound from me in return.'

'I know him,' Cúchulainn said. 'That was Illann, Fergus mac Roich's son.'

'This wound,' the healer said, 'was the work of two warriors.'

'Yes,' Cethern said. 'A pair of them came at me together, with two long shields. They had two tough silver chains and a silver belt each, and two five-pronged spears, banded plain and silver. Each had a collar of silver.'

'I know them,' Cúchulainn said. 'Those were Oll and Oichne, two of Ailill's and Medb's foster-sons. They never go to battle unless they are certain someone will fall at their hands.'

'Then two more warriors set upon me,' Cethern said, 'bright and noble and manly in looks.'

'I know them,' Cúchulainn said. 'Those were Bun and Mecon, Trunk and Root, from the king's most trusted people.'

'The blood is black here,' the healer said. 'They speared through your heart at an angle and made a cross inside you. I can't promise to cure this,' he said, 'but there are a couple of ways I might keep it from carrying you off.'

'And this,' the healer said, 'was the bloody onslaught of two forest kings.'

'Yes,' Cethern said. 'A pair of light-haired warriors set upon me, their faces the size of wooden bowls, one bigger than the other. Yes indeed,' he said, 'each of them pierced the other's point inside me.'

'I know them,' Cúchulainn said. 'Those were two warriors from Medb's great household, Braen and Láréne, "two sons of three lights," the two sons of the forest king.'

'And this,' the healer Fingin said, 'was an attack by three nephews.'

'Yes,' Cethern said. 'Three men, all alike, set upon me.

They had a bronze chain between them, deadly with spikes and spears.'

'Those were the three scabbards of Banba, from Cúroi mac Dáiri's people.'

'This one,' Fingin said, 'was dug by three soldiers.'

'Yes,' Cethern said. 'Three warriors set upon me with war-clubs, wearing three collars of silver round their necks. Each had a handfull of lances and stuck a spear in me, but I stuck him back with it.'

'Those were warriors from Iruath,' Cúchulainn said.

'They pierced you expertly inside the wound,' the healer said. 'They have cut the bloody sinews of your heart. It is rolling around inside you like a ball of wool in an empty bag.'

'Here,' Fingin said, 'I see the work of three furious men.'

'Yes,' Cethern said. 'Three great fat grey-bellied men came at me, discussing my good points as they came.'

'Those were three of Medb's and Ailill's stewards,' Cúchulainn said, 'Scenb and Rann and Fodail — carver, apportioner and server.'

'These three blows were struck in the morning,' Fingin said.

'Yes,' Cethern said. 'Three warriors attacked me, wrapped in black fur cloaks worn bald. Their hooded tunics were covered in stains and they carried three iron cudgels in their hands.'

'Those were the three madmen of Baiscne, three murderous servants of Medb,' Cúchulainn said.

'Two brothers attacked here,' Fingin said.

'Yes,' Cethern said. 'Two great warriors in dark green cloaks set upon me, with curved scallop-edged shields. Each of them had a broad, grey, slender-shafted stabbing-spear in his hand.'

'I know them,' Cúchulainn said, 'Cormac "the king's pillar" and Cormac, Mael Foga's son.'

'Their wounds came close together,' the healer said. 'They got into your gullet and worked there with their deadly javelins.'

'Two brothers struck you in this place,' he said.

'You may be right,' Cethern said. 'A pair of warriors set upon me, one with a head of yellow curls, the other with a head of dark curls. They carried two bright shields graven with gold animals, and two bright-hilted iron swords. Red-embroidered hooded tunics were wrapped about them.'

'I know them,' Cúchulainn said. 'Those were Maine Athramail, the fatherlike, and Maine Máthramail, the motherlike.'

'This is the double wound of a son and a father,' the healer said.

'Yes,' Cethern said. 'Two huge men came at me, their eyes bright as torches, with gold crowns on their heads. They had gold-hilted swords at their waists. Scabbards with tassels of speckled gold hung down to their feet.'

'I know them,' Cúchulainn said. 'Those were Ailill and his son, Maine Cotagaib Uli, who has the likeness of all.'

'Tell me, friend Fingin, what you think of my state.'

'I'll tell you no lie,' Fingin said. 'Don't look to your cows now for calves. If it were only a question of twos or threes . . . But your case is clear — a whole horde has left its tracks in you, and one way or another your life is done.'

Fingin turned his chariot away.

'Your advice is only the same as the others,' Cethern said, and he struck him with his fist and sent him across the chariot's two shafts and smashed the chariot itself.

'That was a wicked blow to give an old man!'

Cúchulainn said. (It is from his word 'Luae'— the blow,
or kick — that the name Ochtur Lui, in Crích Rois, is
taken.) 'Save your kicks for your enemies.'

After this the healer gave him a choice : either to treat
his sickness for a whole year and live out his life's span,
or get enough strength quickly, in three days and three
nights, to fit him to fight his present enemies. He chose
the second course. The healer asked Cúchulainn for bone-
marrow to heal him, and Cúchulainn went out and took
what beasts he could find and made a mash of marrow
out of their bones. From this comes the name Smirommair,
the Bath of Marrow, in Crích Rois.

Cethern slept day and night in the marrow, absorbing
it. He said afterward :

'I have no ribs left. Get me the ribs out of the chariot-
frame.'

'I'll get them for you,' Cúchulainn said.

Then Cethern said :

'If only I had my own weapons, I'd do things they
would talk about for ever.'

'I see something like them coming,' Cúchulainn said.

'What is it?' Cethern said.

'I think I see Finn Bec coming toward us in a chariot —
Eochaid's daughter, your wife.'

The woman came in sight with Cethern's weapons in
the chariot. Cethern took his weapons and made off
toward the armies, with the frame of his chariot bound
around his belly to give him strength. The healer Itholl,
who had lain like a dead man among the bodies of the
other healers, went ahead to warn the Connacht camp.
In their dread, they put Ailill's crown on top of a pillar-
stone, and Cethern attacked the pillar-stone and drove
his sword through it, and his fist after the sword. This
is the origin of the name Lia Toll, the Pierced Stone, in

Crích Rois.

'You have played me false,' Cethern cried. 'I'll give you no rest, now,' he said, 'until one of you puts on this crown of Ailill's.'

He ground them down day and night until one of the Maine placed the crown on his head and came against him in his chariot. Cethern threw his shield at him and it split him and his charioteer open, and cleaved through his horses into the ground. Then the armies closed in on him and he wrought havoc among them until he fell.

Fintan came to avenge his son Cethern, with three times fifty belted and bristling men, all with double-headed spears. They fought seven battles against the enemy, and only Fintan himself and his son Crimthann came out alive, and not one of their followers. Crimthann got separated from his father by a wall of shields and was saved by Ailill out of fear of Fintan, on condition he would fight them no more until he came with Conchobor to the last Battle. Fintan promised friendship to Ailill for giving him back his son.

When Fintan's people and the men of Ireland were found, they had each other caught by the lips and noses in their fanged teeth.

Menn mac Sálchada went against them with thirty bristling men. Twelve of Medb's men fell there and twelve of his own as well. Menn himself was badly wounded and all his followers reddened with blood.

The men of Ireland said:

'It is a red shame for Menn mac Sálchada — his people slaughtered and ruined, and he himself wounded and red with blood.'

Menn was let leave the encampment and no more men were killed. They told him they wouldn't think it any dishonour for him to go back to his home in the watered lands by the Boann river. He went and stayed there. He thought it no dishonour to leave the camp until such time as he was to come with Conchobor to the last Battle.

Cúchulainn told his charioteer to go for help to Rochad mac Faithemain. The charioteer found him and told him to come and help Cúchulainn if his pangs were finished; he said they could steal up on some of the host and destroy them. Rochad came southward with a hundred warriors.

'Scan the plain for us,' Ailill said.

'I see a troop crossing the plain,' the watcher said. 'They have a tender youth among them and they reach up only to his shoulders.'

'Who is that, Fergus?' Ailill said.

'That is Rochad mac Faithemain,' he said, 'coming to help Cúchulainn.'

'Here is what to do,' Ailill said: 'send out a hundred warriors into the middle of the plain with the girl Finnabair in front of them. Send a horseman to tell him that the girl wants to speak alone with him. Then you can get your hands on him and end any danger from his army.'

This was agreed. It happened that Finnabair loved Rochad, for he was the handsomest hero in Ulster at that time, and she had gone to her mother Medb to speak about it.

'I have loved this warrior a long time.' she said. 'He is my true and first and chosen love.'

'If you have so much love for him,' Ailill and Medb said, 'sleep with him tonight and ask him for a truce for our armies until he comes against us with Conchobor on the day of the great Battle.'

Rochad came to meet the horseman, who said:

'I have come to you from Finnabair. Will you talk to her?'

Rochad went alone to talk to her. The troop rushed at him from all sides and grasped him in their arms. So he was captured and his followers fled. Later he was set free on his promise not to fight the armies until the coming of the whole of Ulster. He was offered Finnabair for this and he took her. The girl slept with him that night. Then he returned to Ulster.

The seven kings of Munster were told that Rochad had slept with the girl. One of them said:

'That girl was promised to me, with fifteen hostages as a guarantee, to get me to join this army.'

All seven confessed in turn that she had been promised to them. They came to take vengeance against Ailill's sons who were keeping watch over the armies in Glenn Domain. But Medb rose up against them, and the Galeóin troop of three thousand rose up also, and Ailill and Fergus. Seven hundred died slaughtering each other there in Glenn Domain.

When Finnabair heard that seven hundred men had died because of her deceit, she fell dead of shame. From this comes the place-name Finnabair Slébe, Finnabair in the Mountains.

Then Ilech came against them at Ath Feidli. He was Laegaire Buadach's grand-father; Laegaire was the son of Connad the Yellow-haired, Ilech's son. Ilech had been left under Laegaire's care at Ráith Impail. He came to take vengeance on the army in a decrepit old chariot without covers or cushions. Two old yellow horses pulled the ancient chariot. The whole frame was filled with stones and clods that he flung at everyone who came up to look at him in his nakedness, with his narrow tool and his balls hanging down through the chariot floor. The army kept jeering at the spectacle of the naked man, but Dóchae mac Mágach stopped the rabble at their mocking, and called out to Ilech that he would take his sword and his head from him at the end of that day, if he didn't get out of the army's way.

Ilech saw the mash of marrow. They told him it was made out of Ulster cows' bones. That day he made another marrow-mash beside it, a trench of marrow out of Connachtmen's bones. At nightfall Dóchae cut off Ilech's head and brought it to his grandson Laegaire. He made a pact of friendship with Laegaire and kept the sword.

The armies moved next toward Tailtiu, where three times fifty of Ulster's charioteers attacked them. They killed three times their own number, but they themselves were all killed. Roi Arad, the Battlefield of the Charioteer, is the name of the place: a charioteer and his company fell there on the Táin Bó Cuailnge.

The armies beheld one evening a great stone hurtling upon them from the east, and another like it from the west. The two stones met in the air and fell over the camps of Fergus and Ailill. This playful sport continued until the same hour next day, while the armies sat still with their shields held over their heads to guard against the

blocks of stone, and the plain grew full of stones. This is the origin of Mag Clochair, the Stony Plain. Cúroi mac Dáiri was the cause of this. He had come to help his own people, and stopped at Cotail. Munremur mac Gerrcinn was stopped opposite him at Ard Roich: he had come from Emain Macha to help Cúchulainn. Cúroi knew there was nobody in the armies who could withstand Munremur. It was these two between them that made the sport.

The armies asked them to be still. Munremur and Cúroi made a pact: Cúroi went back to his home and Munremur went to Emain Macha. Munremur didn't come again until the day of the Battle.

While these things were happening, the pangs of the Ulstermen were coming to an end. From Ráith Sualdaim, his house on Murtheimne plain, Sualdam heard how his son Cúchulainn was being harassed.

'Are the heavens rent?' he said. 'Is the sea bursting its bounds? Is the end of the world upon us? Or is that my son crying out as he fights against great odds?'

He went out to his son but Cúchulainn didn't want him there, for if anyone killed him he would have no strength to avenge him.

'Go to the men of Ulster,' Cúchulainn said. 'Tell them to come and fight these armies now. If they don't come soon, they'll never get their revenge.'

His father could see that there was no part of his body bigger than the tip of a rush that hadn't been pierced. In his left hand alone, though his shield here protected it, there were fifty bloody places.

Sualdam went to Emain, and cried out to the men of Ulster:

'Men murdered, women stolen, cattle plundered!'

He gave his first cry from the slope of the enclosure, his second beside the fort and the third cry from the Mound of the Hostages inside Emain itself. Nobody answered. (In Ulster no man spoke before Conchobor, and Conchobor wouldn't speak before the three druids.) Then a druid said:

'Who is robbing and stealing and plundering?'

'Ailill mac Mata,' Sualdam said, 'with the knowledge of Fergus mac Roich. Your people are harassed as far as Dún Sobairche and their cattle and their women and all their herds taken. Cúchulainn kept them out of Murtheimne Plain and Crích Rois; for three winter months now he has fastened his cloak round him with hoops of twigs and kept dry wisps in his joints. He has been wounded so sorely that his joints are coming asunder.'

'This man is annoying the king,' the druid said. 'By rights he ought to suffer death.'

'It would be fitting,' Conchobor said.

'It would,' all the men of Ulster said.

'Still, what Sualdam says is true,' Conchobor said. 'They have been overrunning us from the Monday at summer's end to the Monday at spring's beginning.'

It seemed to Sualdam that they were not doing enough, and he ran out while they were speaking. But he fell over his shield and the scalloped rim cut his head off. His head was brought back on a shield into his house in Emain, where it uttered the same warning again.

'Why all this uproar?' Conchobor said. 'Isn't the sea in front of them still? the sky overhead? the earth under foot? I'll beat them in battle, and bring back every cow to its byre, and every woman and child back home again.'

Conchobor laid his hand upon his son, Finnchad Fer Benn, the Horned Man — so called because of the silver horns he wore — and said:

'Rise up now, Finnchad, and summon Deda to me, from
his bay, and Leamain and Fallach and Fergus's son Illann
from Gabar; Dorlunsa from Imchlár, Derg Imderg the
Red, Fedilmid Cilair Chetaig, Faeladán and Rochad mac
Faithemain from Rígdonn; Lugaid and Lugda; Cathbad
from his bay; the three named Coirpre from Aelai, Laeg
from his causeway and Gemen from his valley; Senoll
Uathach, the Hideous, from Diabal Arda, and Fintan's
son Cethern from Carlaig; Cethern from Eillne, Aurothor,
and Mulach from his fortress; Amargin the royal poet, and
Uathach of the Badb; the great queen at Dún Sobairche;
Ieth and Roth and Fiachna from his mound; Dam Dreimend, Andiaraid and Maine mac Briathrach; Dam Derg,
Mod and Maithes; Irmaithis from Corp Gliath; Gabar from
Laigi Líne, Eochaid from Saimne and Eochaid from Lathar-
nu; Uma mac Remanfisig from Fedan; Munremur mac
Gerrcinn, and Senlobair from Canann Gall; Follamain,
and Lugaid, king of the Fir Bolg, at Laigi Líne; Buadgalach
and Ambuach, and Fergna from Barréne; Aine and Errge
Echbél the horse-lipped; Abra, and Celtchar mac Uthidir
from Lethglas; Laegaire Milbél the honey-mouthed from
his hearth; the three sons of Dromscailt mac Dregamm;
Drenda and Drendas and Cimbe; Cimling and Cimmene
from the slopes of Caba; Fachtna, Sencha's son, in his
rath; Sencha and Senchairthe; Briccir and Bricirne; Breic
and Buan and Bairech; Aengus and Fergus, Léte's sons,
and Aengus of the Fir Bolg; Bruchur and Alamiach of the
old tribes in Slánge, and the three sons of Fiachna from
Cuailnge; Conall Cernach from Midluachair, Connad mac
Morna from Felunt, Cúchulainn mac Sualdaim from
Murtheimne, Amargin from Es Ruaid and Laeg from Léire;
Sálchada's son at Correnna and Cúroi mac Amargin in his
rath; Aengus Fer Benn Uma of the copper horns, and
Ogma Grianainech, whose face is like the sun, from Brecc;

Eo mac nOircne, and Tollchenn from Saithi, and Mogoll
Echbél from the Plain of Ai; Connla Saeb from Uarba,
Laegaire Buadach from Impail; Ailill son of Amargin from
Tailtiu, and Furbaide Fer Benn, the horned one, from
Seil on Inis Plain; Cúscraid Menn the stammerer, the sons
of Lí, and Fingin from Finngabar; Cremath and the hostel
keepers Blai Fichit and Blai Briuga at Fesair; Eogan mac
Durthacht from Fernmag; Dord and Seirid and Serthe,
Oblan from Cuilenn, Cuirther and Liana from Eith Benne,
Fernel and Finnchad from Sliab Betha, Talgobain from
Bernas, and Menn mac Sálchada from Dulo Plain; Iroll
from Blarigi, Tibraide mac Ailcotha; Iala the ravager from
Dobla Plain, Rus mac Ailcotha, Maine mac Cruim, Nin-
nech mac Cruinn, Dipsemilid and Mál mac Rochrad;
Muinne, Munremur's son, Fiatach Ferndoirre son of
Dubthach, and Muirne Menn.'

Finnchad found that his task was easy, for all the chief-
tains in Conchobor's province had been waiting for
Conchobor to move. They had gathered around Emain
east and north and west and entered Emain Macha in time
for Conchobor's wakening.

They moved out of Emain southward to look for the
armies. The first stage of their march was from Emain to
Iraird Cuilenn.

'Why are you stopping here?' Conchobor said.

'We are waiting for your sons,' they said. 'They are
gone with thirty others to Temair to get Erc son of Coirpre
Niafer and Fedelm Noichride. We're not leaving here until
their two troops of three thousand arrive.'

'I can't wait until the men of Ireland discover I have
risen from my pangs,' Conchobor said.

Conchobor and Celtchar went ahead with three times
fifty chariots and came back with eight score men's heads
from Airthir Midi Ford, in East Meath. The ford is known

ever since as Ath Féne, Warriors' Ford. These eight score
warriors had been keeping watch for the armies. They
had eight score women with them, their share of the
plunder. When Conchobor and Celtchar brought the
heads back to the camp, Celtchar said to Conchobor:

> 'Slingshots reddened
>> by a terrible king
> proud past compare
>> sinews split
> limp with horror
>> in a hundred branches
> ground given up
>> of fourhorse chariots thirty
> of the host's hard steeds a hundred
>> two hundred druids to lead us
> a solid man not lacking
>> at Conchobor's back
> prepare for the battle
>> let the warriors wake
> the battle breaks out
>> at Gáirech and Irgairech.'

Others say it was Cúscraid the stammerer of Macha,
Conchobor's son, who chanted this the night before the
battle, just after Laegaire Buadach had made the chant
beginning: 'Rise kings of Macha,' and that he chanted it
in the eastern camp.

During the night Dubthach Dael of Ulster, the black-
tongued, dreamed of the armies at Gáirech and Irgairech,
and spoke in his sleep:

'Monstrous morning
          monstrous season
hosts in turmoil
          kings cast down
necks broken
          a red sun
three hosts crushed
          by the host of Ulster
about Conchobor
          woman struggle
herds driven
          the morning following
heroes felled
          hounds cut down
horses mangled
          tunics torn
the earth drinking
          spilt blood
of gathered hordes.'

This upset them in their sleep. The Nemain brought
confusion on the armies and a hundred of their number
fell dead. Silence fell again, until Cormac Connlongas
(or some say Ailill mac Mata) was heard chanting in the
western camp:
          'Ailill's hours!
          A great truce
                    the truce at Cuillenn
          a great plot
                    the plot at Delind
          great herds of horses
                    the herds at Assal
          a great plague
                    at Tuath Bressi.'

XIII   THE COMPANIES ADVANCE

DURING this time, the Connacht army took counsel with Ailill and Medb and Fergus. They decided to send scouts to see if the men of Ulster had reached the plain. Ailill said:

'Mac Roth, go and see if they are all here on the plain of Meath. If they are not here yet, I have got clear away with their goods and herds. They can look for fight as much as they like now, I'm not waiting here for them any longer.'

Mac Roth went off and scanned and scoured the plain, and hurried back to Ailill and Medb and Fergus. When he first looked out from the Sliab Fuait road he had seen all the wild animals leaving the forest and coming out over the plain.

'I looked a second time,' Mac Roth said. 'I studied the plain before me and saw a dense fog filling the valleys and hollows, so that the high places in between looked like islands in a lake. I made out sparks of fire through the thick fog, and a world of different colours, of all kinds. Then I saw flashes of lightning, with uproar and thunder. Though there is only a light breeze out today, a great wind came that flung me down on my back and all but swept the hair from my head.'

'What is this, Fergus, do you think?' Ailill said.

'I know well what it is. The men of Ulster have risen from their pangs. It is they who entered the forest, great heroes thronging in might and violence; and they who shook the forest and sent the wild animals fleeing onto the plain. The dense fog you saw filling the hollows, that was the breath of those fierce men filling the valley until the hills in between looked like islands in a lake. The flashes of lightning and the sparks of fire and all those colours you saw, Mac Roth,' Fergus said, 'those were the warriors' eyes, so bright you thought they were sparks of fire. The thunder and thudding and turmoil you heard, that is the humming of their blades and their ivory-hilted swords, the uproar of arms, the clattering of chariots — horse-hooves hammering, fierce chariot-fighters — the outcry of an army: the sound of warriors, the anger and fury and ferocity of the brave as they rage toward the battle. They think they will never reach it, their angry spirit is so high,' Fergus said.

'Let them come,' Ailill said. 'We have warriors to meet them.'

'You'll need them,' Fergus said. 'No one in all Ireland, or the western world from Greece and Scythia westward to the Orkney Islands and the Pillars of Hercules, as far as the tower of Breogan and the Islands of Gades, can withstand the men of Ulster when their fury is roused.'

Mac Roth went off once more to see how the men of Ulster were coming. He went up to their encampment on the smooth plain of Slemain Midi. Then he came to Ailill and Medb and Fergus and gave them this news:

'A mighty great force, fierce and ferocious, came to the hill at Slemain Midi,' Mac Roth said, 'a full troop of three thousand, I would say. They tore their clothes off straight away and dug a mound of sods where their leader was to sit. He was fair and graceful and tall, a choice royal figure out before his company, handsome and slender. He had light yellow hair cut and curled neatly and reaching down in waves to the shallow between his shoulders. He wore a purple pleated tunic wrapped around him. A rich brooch of red-gold fastened the cloak at his breast. His eyes were very grey and gentle, his face bright and blushing, the brow broad, the jaw narrow. He had a forked and wreathed gold beard. He wore a white, red-embroidered hooded tunic and carried a gold-hilted sword reaching to his shoulders, with a bright shield graven with gold animals. He held in his hand the slender shaft of a broad grey stabbing-spear. The finest of the world's princes in figure and dress and fury and following, he advanced with looks of strife, terror, triumph, rage and fierce dignity.'

'Another company came,' Mac Roth said, 'second only to the first in numbers and discipline and dress and terrible fierceness. A fair young hero headed this company, with a green cloak wrapped around him, fastened at his shoulder with a gold brooch. His hair was curled and yellow. He wore at his left an ivory-hilted sword, the hilt cut from a boar's tusk. A bordered tunic covered him to the knee. He carried a scallop-edged, death-dealing shield and a great spear in his hand like a palace-torch with silver rings running one way along the shaft as far as the tip, then running back to the grip. That company settled at the left hand of the leader of the first company. They squatted with their knees on the earth and their shield-rims at their chins. I thought I heard a stammer in the speech of the great grim champion who led that company.'

'Another company came,' Mac Roth said. 'It looked more than a full troop of three thousand. A wild and wilful man went before them, broad-headed and fair-featured. He had brown curly hair and a long thin forked beard. He wore a dark-grey fringed cloak wrapped around him, caught on his breast by a leaf-shaped pin of light gold. A white hooded tunic covered him to the knee. He carried a hero's shield graven with animals, a naked sword with a bright silver grip at his waist and a five-pronged spear in his hand. He sat down and faced the leader of the first company.'

'Who are these, Fergus?' Ailill said.

'I know these companies well,' Fergus said. 'Conchobor, king of a province of Ireland, is the one who settled himself on the mound of sods. Sencha mac Ailella, the most eloquent man in Ulster, is the one who sat facing him. Cúscraid Menn Macha, the stammerer, Conchobor's son, is the one who sat at his father's hand. The spear in his hand always plays like that just before a victory: the rings won't run round it at any other time. That was a great group for finding fight and serving out wounds,' Fergus said.

'They'll find what they want here,' Medb said.

'I swear to my people's god,' Fergus said, 'the army wasn't raised in Ireland yet that can resist the men of Ulster when they are provoked.'

'Another company came,' Mac Roth said, 'a troop of three thousand and more, with a great swarthy fiery-faced champion at its head, awesome and terrible. His dark brown hair lay flat on his forehead. He carried a curved scallop-edged shield, with a five-pronged spear in his hand, a forked javelin at his side, and a cruel sword slung behind him. A purple cloak was wrapped around him with a gold brooch at the shoulder. A white hooded tunic covered him to the knee.'

'Who is that, Fergus?' Ailill said.

'The beginner of battle.' Fergus said, 'a man created for war. He falls on his enemies like a doom: Eogan mac Durthacht, king of Fernmag.'

'Another great grim company came to the hill at

Slemain Midi,' Mac Roth said, 'with their cloaks thrown
back behind them. Dark and steady they came to the hill
bringing great dread and terror, I tell you no lie. The
clash of their weapons was awful as they marched. A
great fearsome champion with a fleshy head was their
leader, with sparse grey hair and big yellow eyes. He was
wrapped in a yellow cloak with a white border. A scallop-
edged, death-dealing shield hung at his side. He carried a
broad-bladed javelin and a long spear with a blood-stained
shaft. Next to it in his hand was another javelin, with the
blood of enemies on its blade. A big murderous sword
hung at his shoulders.'

'Who is that, Fergus?' Ailill said.

'A warrior who has never shirked the warlike fray.
That is Laegaire Buadach, the victorious, son of Connad
son of Ilech from Impail in the north,' Fergus said.

'Another great company came to the hill in Slemain
Midi,' Mac Roth said, 'with a pleasant, fat, thick-necked
warrior at their head. His hair was black and curled, his
face flushed, his grey eyes bright. He wore a noble brown-
ish cloak about him held by a bright silver brooch. He
carried a black shield with a knob of bronze and a shimmering spear in his hand, set with eyes. A red-embroidered
braided tunic covered him and an ivory-hilted sword
hung out over his clothes.'

'Who is that, Fergus?' Ailill said.

'The first in the fray: he advances like a devouring
sea-wave over a little stream. A man of three cries. He
falls on his enemies like a bitter doom,' Fergus said.
'Munremur mac Gerrcinn, from Moduirn in the north.'

'Another great company came to the hill at Slemain
Midi,' Mac Roth said, 'a fine numerous and handsome
company, well dressed and disciplined. They hurried
fiercely to the hill. They shook the armies with the clash

of their weapons as they advanced. A pleasant proud champion came at their head, the most marvellous among men for his hair and eyes and grim aspect, for apparel, bearing, voice, paleness and proud lofty good looks, for weapons and skill and style, for equipment, apt feats, learning, distinction and breeding.'

'As you describe him,' Fergus said, 'that was the bright flame, the fair Fedilmid, an overwhelming storm wave, coming in warrior's rage and irresistible might, full of triumph, from the destruction of his enemies in other lands: Fedilmid Cilair Chetaig.'

'Another company came to the hill at Slemain Midi,' Mac Roth said, 'a full warlike troop of three thousand at the least count, with a great upright sallow warrior bravely at their head. His hair was black and curly, his dull-brown eyes scornful and large. A harsh, firm, bull-like man. He wore a grey cloak around him, held at the shoulder with a silver pin, and a white hooded tunic. He carried a sword at his thigh, and a red shield with a knob of tough silver. He held a broad, triple-rivetted blade in his hand.'

'Who is that, Fergus?' Ailill said.

'Connad mac Morna, coming from Callann in an angry glow, bold in battle, the winner of wars,' Fergus said.

'Another company came to the hill at Slemain Midi,' Mac Roth said, 'big as an army in its advance. Seldom will you find a champion of better style and bearing than the leader at the head of that company. His red-gold hair was close-cropped, his flushed face fine and well formed— the jaw narrow, the brow broad — with fine red lips and shining pearls for teeth. His voice rang clear as he lifted his fine, flushed, well-formed face, the most marvellous among men. He wore a purple cloak wrapped around him with an elaborate gold brooch on his white breast. He carried at his left a curved shield with a knob of silver, graven with all kinds of coloured animals. He held in his hand a long bleak-bladed javelin and a keen quick spear. A sword of gold with a gold hilt hung at his back. A red-embroidered hooded tunic wrapped him round.'

'Who is that, Fergus?' Ailill said.

' I know him well,' Fergus said, 'half an army in himself, a barrier in battle, a ravening mastiff, Rochad mac Faithemain from Rígdonn, your own son-in-law that took your daughter Finnabair.'

'Another company came there to the hill at Slemain Midi,' Mac Roth said, 'with a burly, thick-thighed, brawny-calved hero at their head. His legs and arms were each as thick as a man. From head to foot he was a man indeed,' he said. 'His hair was black, his face scarred and fiery, with scornful, blazing bloodshot eyes : a sprightly splendid man in every way, horrible and grim. The style and dress and weapons of his warriors made a marvellous spectacle as he came among them in triumph — a hero full of warlike deeds and wilful dignity as he goes against great odds to crush the foe in his anger, scorning fair fight, or as he travels unprotected through hostile lands. Steadily they advanced on Slemain Midi.'

'That was a flood of skill and courage,' Fergus said, 'a flood of hot blood, vigour, power and pride — a force to hold armies together : my own foster-brother, Fergus mac Léte, king of Líne, the battle-crest of the north of Ireland.'

'Another great grim company came to the hill at Slemain Midi,' Mac Roth said, 'in weird apparel, bringing strife before them, with a fine fair hero at their head, magnificent in every way — for his hair and eyes and pallor, his stature, structure and ferocity. He wore five chains of gold, a white hooded tunic, and a green cloak wrapped around him and fastened at the shoulder with a gold brooch. He held a spear like a palace pillar in his hand. A gold-hilted sword hung at his shoulder.'

'That was a battle-hungry hero, very quick to wrath,' Fergus said: 'Amargin the son of the smith Ecet Salach, the grimy one, from Buais in the north.'

'Another company came to the hill at Slemain Midi,' Mac Roth said, 'overwhelming, fiery in splendour, spiky sharp, their numbers legion, a rock-mass, full of strength, doom in battle, a quick thunder. A terrible hero led that company with harsh looks: big-bellied, big-nosed, thick-lipped, his hair tough and grizzled, his limbs red. He wore a rough woven tunic and a dark cloak about him with an iron spike fastening the cloak. He carried a curved scallop-edged shield and a great grey javelin in his hand, with thirty rivets. A sword that was tempered seven times hung at his shoulders. The whole army rose up to meet him. Troop after troop of them fell into disorder as he proceeded to the hill.'

'That was the topmost glory coming,' Fergus said, 'half an army in himself, he fights so fiercely — a stormy ocean wave breaking over barriers, Celtchar mac Uthidir from Dún Lethglaise in the north.'

'Another company came there to the hill at Slemain Midi,' Mac Roth said. 'At its head was a warrior all in white, his hair and eyelashes and beard all fair and his clothing white. He carried a shield with a knob of gold on it and an ivory-hilted sword in his hand, with a broad, pitted stabbing-spear. He advanced on his way like a high hero.'

'A most cherished, powerful and death-dealing bear,' Fergus said, 'murderous as a bear to the enemy, a man-crusher, the fair and righteous Feradach Finn Fechtnach from the wood at Sliab Fuait in the north.'

'Another company came to the hill at Slemain Midi,'
Mac Roth said, 'with a terrible warrior out in front. He
had a big belly and thick horse's lips. His hair was dark
and curly and he had only one eye. His head was broad
and his hand long. A black cloak swung about him,
fastened with a disk of tin. He carried a dark-grey shield
at his left, and a broad stabbing-spear, banded at the neck,
in his right hand. A long sword hung at his shoulders.'

'A ravening, red-clawed lion,' Fergus said, 'sharp and
fearful and busy in battle, not to be withstood as he rages
on the earth: Errge Echbél, the horse-lipped, from Brí
Errgi in the north,' Fergus said.

'Another company came there to the hill at Slemain
Midi,' Mac Roth said, 'with two tender heroes at their
head — alike in looks, with heads of yellow hair and two
bright shields graven with silver animals. They were the
same age. They raised their feet and set them down
together, neither out of step with the other.'

'Two heroes, two pure flames, two battle-spikes,' Fergus
said, 'two champions and pillars of the fray; two dragons,
two fires, two war-like battle champions; the two props
and spoiled pets of Ulster and its king.'

'Who are they, Fergus?' Ailill said.

'Fiachna and Fiacha, two sons of Conchobor mac Nesa,
the two dear darlings of the north of Ireland,' Fergus said.

'Another company came to the hill at Slemain Midi,'
Mac Roth said. 'Three noble and fiery champions came at
their head, with faces flushed, all three with hair gold-
yellow and cropped. Three cloaks of the same colour
were wrapped about them, fastened at the shoulders with
gold pins. They wore red-embroidered sleeved tunics. They
carried three similar shields. Three gold-hilted swords
hung at their shoulders, and three broad grey spears were

in their hands. All three of an equal age.'

'Three fiery torches from Cuib and Midluachair who have done great deeds, three princes of Roth, three hardened soldiers from east of Sliab Fuait,' Fergus said. 'Those were Fiachna's three sons, Rus and Dáire and Imchath, who have come to recover the bull.'

'Another company came to the hill at Slemain Midi,' Mac Roth said, 'a furious lively man at its head, with hot heroic eyes. He wore a speckled cloak with a silver disk to hold it. He carried a grey shield on his left and a silver-hilted sword by his side. He held in his wrathful right hand a javelin shaped well for subtle thrusts. A white hooded tunic covered him to the knee. The company about him was red with blood, and he himself marked with blood.'

'That was a brave and pitiless one,' Fergus said, 'a gashing beast, a wild boar in battle, a raving bull, the conqueror from Baile and holder of the gap, the torch of battle from Colptha, the protector of the border of the north of Ireland : Menn mac Sálchada from Corann. He has come to avenge his wounds,' Fergus said.

'Another company came to the hill at Slemain Midi,' Mac Roth said, 'spirited and eager, with a great long-cheeked sallow warrior at their head. He had dark curling hair and wore a fine red woollen cloak and a handsome tunic. A gold pin held the cloak at his shoulder. He wore at his left a sword of great beauty with a hilt of bright silver. He carried a red shield and a grey broad-bladed stabbing-spear in his hand, beautifully worked and set onto its shaft of ash.'

'That was a man of three hard strokes,' Fergus said, 'a man of three roads and highways and byways, a man with three qualities and three cries, who breaks foreign enemies in battle — Fergna mac Finnchaime, from Corann.'

'Another company came to the hill at Slemain Midi,' Mac Roth said. 'It seemed greater than a troop of three thousand, with a white-breasted, well-favoured warrior at its head, who looked like Ailill in size and handsomeness and apparel. He wore a gold crown on his head, and a red-embroidered tunic. A cloak of great beauty wrapped him round, fastened on the breast with a gold brooch. He carried a gold-rimmed, death-dealing shield and a spear like the pillar of a palace. A gold-hilted sword hung at his shoulder.'

'Like the sea against a stream he comes,' Fergus said, 'a fiery blaze, in irresistible fury against his foes, Furbaide Fer Benn, the horned man.'

'Another company came to the hill at Slemain Midi,' Mac Roth said, 'heroic, numberless, with strange garments, not like the other companies. Their clothes and all their outfit and weapons were remarkable as they marched. This company was a great angry army in itself, with a flushed freckled boy at its head, the most marvellous of men by his looks. He carried on his arm to the glorious battle a gold-rimmed and gold-inlaid shield with a white knob, and a light sharp javelin shimmering in his hand. He wore a red-embroidered white hooded tunic and a purple fringed cloak wrapped round him, held at the breast with a silver pin. A gold-hilted sword hung out over his clothes.'

At that, Fergus was silent.

'I don't know anyone like that boy in Ulster,' Fergus said, 'unless these are the men of Temair gathered about the fine and noble well-favoured Erc, son of Coirpre Niafer and Conchobor's daughter. Coirpre and Conchobor are not friendly toward each other, and this boy may have travelled to help his grandfather without asking his father's leave. You will lose the battle on account of this boy,' he said. 'He knows no fear or terror, and when he presses into the midst of your forces the fighting-men of Ulster will raise a manly shout and hack through the fray to save the little calf of their hearts. Seeing the boy in such terrible turmoil they will all fill up with sudden affection and hack a path through the battle. Then the humming of Conchobor's sword will be heard like a mastiff growling as he comes to save the boy. Conchobor will throw up three mounds of men around the battlefield in the search for his little grandson. And, full of family feeling, the inflamed fighting-men of Ulster will fall on your countless army,' Fergus said.

'I am tired,' Mac Roth said, 'describing all I have seen. But there is something more to say.'

'You have said enough,' Fergus said.

'Nevertheless,' Mac Roth said, 'Conall Cernach and his great company haven't come. Conchobor's three sons and their three troops of three thousand haven't come. Cúchulainn, wounded in the unequal struggle, hasn't come. Many hundreds, many thousands, have reached the Ulster camp. Many heroes and champions and warriors have hurried there to the gathering. But more companies still were on their way there as I left,' Mac Roth said. 'My eye travelled from Ferdia's Ford to Slemain Midi and fell on men and horses instead of hills and slopes.'

'You have certainly seen a man of some following,' Fergus said.

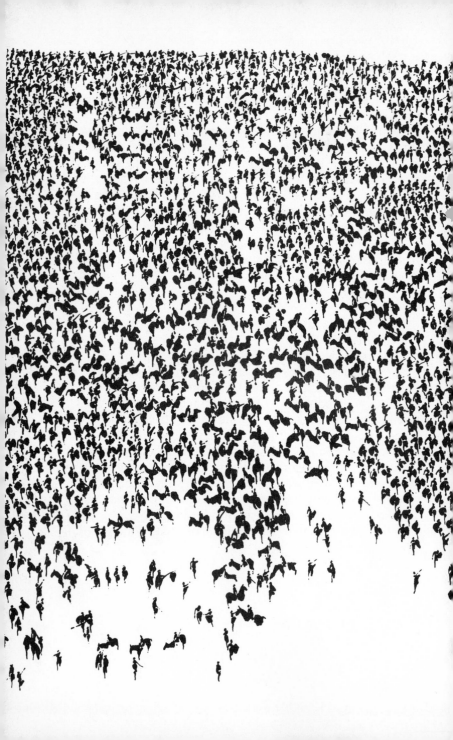

CONCHOBOR came with his armies, and spoke to Ailill about a truce until sunrise. Ailill agreed for the men of Ireland and the exiles, and Conchobor agreed for the men of Ulster. Conchobor's tents were pitched and he settled in his camp surrounded by his followers. The men of Ulster were settled before sunset. The ground between the armies lay bare.

In the half light between the two camps, the Morrígan spoke:

> 'Ravens gnawing
>> men's necks
> blood spurting
>> in the fierce fray
> hacked flesh
>> battle madness
> blades in bodies
>> acts of war
> after the cloaked one's
>> hero heat
> in man's shape
>> he shakes to pieces
> the men of Cruachan
>> with hacking blows
> war is waged
>> each trampling each.
> Hail Ulster!
>> Woe men of Ireland!
> Woe to Ulster!
>> Hail men of Ireland!'

This last ('Woe to Ulster') she said in Connachtmen's ears only, to hide the truth from them. That same night Nét's wives, Nemain and the Badb, called out to the men of Ireland near the field at Gáirech and Irgairech, and a hundred warriors died of fright. It was a bad night for them.

Ailill mac Mata chanted on the eve of the battle, saying:

'Rise up, Traigthrén, swift-footed. Summon for me the three called Conaire from Sliab Mis; the three fair ones called Les, in Luachair; the three called Meid from Corpthe Loste; the three named Buidir from the river Buas; the three called Badb from the river Buaidnech; the three called Buaideltach from the river Berba; the three Muredachs from Marga; the three Laegaires from Lec Derg; the three Suibnes from the river Siuir; the three Echtachs from Ane; the three called Dael from Eirc; the three called Damach from Derg Derc; the three called Bratruad from Loch Rí; the three named Nelleth from Loch Eirne; the three named Bresal from Bodg; the three named Amalgad from Ai; the three Fiachras from Nemain Wood; the three Nechtas from Muiresc Plain; the three famous sons from Es Ruaid; the three Ruirechs from Aigle; the three Bruchurs from the river Febrad; the three Conalls from Collamair; the three named Féic from Finnabair; the three Coirpres from Cliu; the three named Maine Milscothach; the three Descertachs from Drompa; the three Fintans from Femen Plain; the three Rathachs from Raigne Plain; the three Eterscéls from Eterbán; the three Guaires from Gabail; and the three named Aed from Aidne.'

These men, in groups of three, were all the men of Ireland that survived the former slaughter by Cúchulainn.

At this time Cúchulainn was lying nearby at Fedan Chollna. The landowners there visited him every day and supplied him with food at night. West of Ferdia's Ford he had killed no one.

'A small herd of animals has strayed from the western camp over toward our camp in the east, and some servants are coming out after them to bring them back,' the charioteer said to Cúchulainn. 'Now some of our servants from the eastern camp are coming out to take them from them.'

'The servants will start fighting,' Cúchulainn said, 'and the animals will go wandering over the plain while everybody goes to help the servants.'

This is what happened.

'How are the Ulster servants fighting?' Cúchulainn said.

'Like true men,' the charioteer said.

'Their honour would make them die for the sake of their herds,' Cúchulainn said. 'What is happening now?'

'The beardless boys have joined the fight,' the charioteer said.

'Has the light of the sun touched the clouds yet?' Cúchulainn said.

'Not yet,' the charioteer said.

'If only I had the strength to join them!' Cúchulainn said.

'There will be enough slaughter today without that,' the charioteer said. 'Now it is sunrise. The better-born people are joining the battle. The kings haven't come yet; they are still asleep.'

It was at sunrise that Fachtna said (or some say it was Conchobor who chanted it in his sleep):

> 'Rise kings of Macha
> > modest people
> of mighty acts
> > blades are battering
> battle raging
> > the earth torn up

shields beaten
    arms weary
herds bellowing
    in the rightful fight
battle ranks trampled
    underfoot
lords and princes
    lead in battle
or end in blood
    a forest of men
where they march and fall
    bitter blood drained
hearts of queens
    filled with grief
the dire advance
    grass soaked with blood
where they stand and fall
    rise kings of Macha.'

'Who chanted that?' everyone said.
'Conchobor mac Nesa,' some said.
'Fachtna,' others said.
'Sleep, sleep — but with your sentries watchful.'

Laegaire Buadach said:
    'Rise kings of Macha
        look to your cattle
    guard your plunder
        drive Connacht's force
    from Uisnech hill
        men's flanks in danger
    sinews on fire
        he will fell the world
    on the field of Gáirech.'

'Who chanted that?' everyone said.

'Laegaire Buadach mac Connaid Buidi meic Ilech. Sleep, sleep — but with your sentries watchful.'

'Wait a little longer,' Conchobor said, 'until the sun has lit all the hollows and hills of Ireland.'

Then Cúchulainn, from the east. saw the Connacht kings setting their crowns on their heads and coming to relieve their companies. He told his charioteer to rouse the men of Ulster. The charioteer spoke (though some say it was the poet Amargin mac Ecit):

> 'Rise kings of Macha
> > modest people
> of mighty acts
> > the Badb covets
> the cattle of Impail
> > heart's gore poured out
> strife fills men's veins
> > to feed brave acts
> panic flight
> > heart's gore on the ground
> the battle din dies
> > is there none like Cúchulainn
> to work Macha's will
> > for Cuailnge's cattle
> rise early now.'

'I have woken them,' the charioteer said. 'They are rushing naked to the battle, with nothing but their weapons. Those that were facing to the east have dashed out through the backs of their tents!'

'Necessity is a great spur,' Cúchulainn said.

Then he said:

'Friend Laeg, how are the men of Ulster doing in the battle?'

'Like true men,' the charioteer said. 'They are fighting closely. Conall Cernach's charioteer En and I could mount our chariots now and drive over the armies from one wing to the other, and find no place for a hoof or a wheel-rim to sink through.'

'It has the making of a great battle,' Cúchulainn said. 'Tell me everything that happens, leave out nothing.'

'I'll do the best I can,' the charioteer said.

'The warriors from the west have reached the eastern battle-line and broken through,' he said presently. 'Now the same number from the east have broken through the western battle-line.'

'Alas!' Cúchulainn said. 'You would see me attacking there with the rest of them if I had my health.'

The men of Ireland. in their groups of three, advanced to the ford nearer the army — a rare spectacle as they marched to the battle at Gáirech and Irgairech. The nine chariot-fighters from Iruath advanced with them, the three on foot out in front as swift as those in the chariots. But Medb held them back from the battle, to pluck Ailill from the fray if their army was beaten or to kill Conchobor if they won.

Now the charioteer told Cúchulainn that Ailill and Medb were asking Fergus to join the fight, reproaching him with all they had done for him during his exile.

'If only I had my sword,' Fergus said, 'I'd send men's severed heads toppling thicker than hailstones over their shields into the mud. It would be like a king's horses churning up the ground. I swear by my people's god.' he said, 'I'd heap up men's hacked jawbones on men's necks,

men's necks on men's shoulders, their arms on their
elbows, with elbows on wrists, wrists on fists, fists on
fingers, fingers on nails, nails on skulls, skulls on trunks,
trunks on thighs, thighs on knees, knees on calves, calves
on feet, feet on toes and toes on nails! I'd send necks
buzzing through the air like bees humming on a fine day!'
    Ailill said to his charioteer:
'Bring me that flesh-piercing sword. I swear by my
people's god, if its bloom has faded since the day I gave
it to you on that hillside in the land of Ulster, not all of
Ireland will save you from me.'
    They brought Fergus's sword and Ailill said:
                  'Now take your sword
                         lay Ireland low
                  but spare us at Gáirech
                         mighty man among boys
                  if all is true
                         not upon us
                  be your wrath wrought
                         let the rage rise
                  against Ulster's heroes
                         at dawn on Gáirech
                  in the red morning
                         we'll see by the sods.'

Fergus said:
                  'Bitter blade welcome
                         Léte's sword
                  bearer of quick
                         Badb horror
                  no longer hidden
                         you have come to avenge
                  a troop of warriors
                         sinews smashed

> heads toppled
>> this sword no longer
> in a sovereign's keeping
>> tales to be told
> my sword will not
>> deal death on you
> but do you glory
>> before the men of Ireland.

'It would be a shame if you were to fall on this glutted field of battle,' Fergus said to Ailill.

Fergus seized his weapons and went into battle. With the sword held in his two hands, he carved a gap of a hundred men in the ranks. A hundred Ulster warriors died by his sword in the first onslaught. He came on Conall Cernach.

'You rage very hard at your kith and kin,' Conall Cernach said, 'for the sake of a whore's backside.'

Then Medb took up her weapons and hurried into battle. Three times she drove all before her until she was turned back by a wall of javelins.

'Who is forcing the battle against us from the north?' Conchobor said to those around him. 'Hold the fighting here and I'll go and find him.'

'We will hold out,' the warriors said, 'until the earth gives under us, or until the heavens fall on us and make us give way.'

Conchobor sought out Fergus and raised his shield against him — the shield Ochain, the Ear of Beauty, with its four gold horns and four coverings of gold. Fergus struck it three blows but couldn't budge even the rim of the shield enough to touch Conchobor's head.

'What man of Ulster holds that shield?' Fergus said.

'A better man than you,' Conchobor said, 'one who drove you out to live in exile with the wild dogs and foxes; one who'll stop you with his battle-deeds today before all the men of Ireland.'

With that, Fergus raised his sword for a vengeful two-handed stroke at Conchobor. As the point touched the ground behind him, Cormac Connlongas flung his arms around him and caught his two hands at the wrist.

'Harshly, harshly, friend Fergus,' Cormac said. 'That would be mean and shameful, and spoil friendships. These wicked blows will cheapen your enmity and break your pacts.'

'Then where am I to strike?' Fergus cried.

'Turn your hand aside. Strike out anywhere. Strike crosswise at those three hills. But remember that Ulster's honour was never thrown away, and never will be unless you do it today. Leave us, Conchobor,' Cormac said to his father. 'This man will pour his rage on Ulstermen no more.'

So Fergus turned aside and struck at the hills. With three strokes he levelled the three bald-topped hills of Meath.

Cúchulainn heard the blows Fergus dealt at the hills, and at Conchobor's shield.

'Who struck those terrible blows in the distance?' he said. 'Blood blocks my heart — battle madness tears! Undo these twigs quickly.'

'Fergus mac Roich, the brave, a man among men, struck them,' Laeg answered; 'Fergus mac Roich, in bloodshed and mounting glory — with the sword that was hidden in the chariot-shaft. The great fight has touched Ochain, our master Conchobor's shield.'

'Loosen the hazel twigs. Quickly!' Cúchulainn said. 'Play of swords — men smothered in blood — bodies swallowed up!'

The wisps of rushes sprang up on high like larks, the bindings of hazel-twigs sprang away from him as far as Mag Tuag, the plain of the hazel-bands, in Connacht, and he ran about this way and that. His wounds opened afresh: Medb had sent two handmaids to lament over him and make his wounds open again, telling him how Fergus was fallen and Ulster broken in battle while he was kept from the fight. But he smashed their heads together, so that each was stained grey from the other's brains. The warp-spasm seized him and they put the twenty-seven skin-tunics around him, all strings and straps, that he wore going into battle. He took his whole chariot on his back, the frame and the two rimmed wheels. Then he rushed toward the battle and circled around looking for Fergus.

'Come here, friend Fergus!' Cúchulainn cried, three times before he was answered. 'I swear by Ulster's god,' he said, 'I'll churn you up like foam churned in a pool! I'll stand up over you like a cat's tail erect! I'll batter you as easily as a loving woman slaps her son!'

'What man in Ireland talks to me like that?' Fergus said.

'Cúchulainn, the son of Sualdam and Conchobor's sister,' Cúchulainn said. 'Give way before me.'

'I promised to do that,' Fergus said.

'It has fallen due,' Cúchulainn said.

'Very well,' Fergus said, 'you ran from me once, and now you are riddled with wounds.'

Fergus went off with his troop of three thousand. The men of Galeóin and the men of Munster went away as well. They left Medb and Ailill to the battle, with their seven sons and their nine troops of three thousand men.

When Cúchulainn joined the battle it was noon. The
sun had reached the tresses of the wood when he smashed
their last company. Nothing was left of his chariot but a
handful of ribs out of the frame and a handful of spokes
from the wheel.

Medb had set up a shelter of shields to guard the rear of
the men of Ireland. She had sent off the Brown Bull of
Cuailnge to Cruachan by a roundabout road, with fifty
of his heifers and eight messengers so that, whoever
escaped, the Brown Bull of Cuailnge would be got safely
away, as she had sworn.

Then Medb got her gush of blood.

'Fergus,' she said, 'take over the shelter of shields at
the rear of the men of Ireland until I relieve myself.'

'By god,' Fergus said, 'you have picked a bad time for
this.'

'I can't help it,' Medb said. 'I'll die if I can't do it.'

So Fergus took over the shelter of shields at the rear
of the men of Ireland and Medb relieved herself. It dug
three great channels, each big enough to take a household.
The place is called Fual Medba, Medb's Foul Place, ever
since. Cúchulainn found her like this, but he held his
hand. He wouldn't strike her from behind.

'Spare me,' Medb said.

'If I killed you dead,' Cúchulainn said, 'it would only
be right.'

But he spared her, not being a killer of women. He

watched them all the way westward until they passed
Ath Luain, and there he stopped. He struck three blows of
his sword at the stone hills nearby. The Bald-topped Hills
is their name now, at Ath Luain, in answer to the three
Bald-topped Hills in Meath.

The battle was over.

Medb said to Fergus:

'We have had shame and shambles here today, Fergus.'

'We followed the rump of a misguiding woman,' Fergus
said. 'It is the usual thing for a herd led by a mare to be
strayed and destroyed.'

They took the bull away on the day after the battle. On
Ai Plain, at Tarbga — the place of bull-grief or bull-strife:
the hill originally called Roi Dedonn — he met the bull
Finnbennach, the White-Horned. Everyone who had
escaped the battle stopped what he was doing, to see
the two bulls fight together.

The men of Ireland asked who should judge between
the bulls. They agreed it should be Bricriu mac Carbad,
because he favoured his friend no more than his enemy.
So he was brought to a gap between the bulls to judge
them. But the two bulls trampled across him as they
struggled, and killed him. Such was Bricriu's death.

The Brown Bull of Cuailnge planted a hoof on the other
bull's horn. All day until nightfall he wouldn't draw the
hoof back toward him. Fergus chided him and took a
stick to his flank.

'It would look bad,' Fergus said, 'to get this quarrelsome
old calf so far, only to have him throw away the honour
of his kind. Men have died on both sides because of you.'

At that, the bull jerked back his hoof. His leg broke, but the other bull's horn was sent flying to the mountain nearby. It is called Sliab nAdarca, the Mountain of the Horn, ever since. Then the bulls fought each other for a long time. Night fell upon the men of Ireland and they could only hear the uproar and fury in the darkness. That night the bulls circled the whole of Ireland. When morning came, the men of Ireland saw the Donn Cuailnge coming westward past Cruachan with the mangled remains of Finnbennach hanging from his horns.

He brandished them before him all that day, and at nightfall entered the lake near Cruachan. He came out with Finnbennach's loins and shoulderblade and liver on his horns. The armies went to kill him, but Fergus stopped them and let him go anywhere he liked. He headed toward his own land. He stopped to drink in Finnlethe on the way. He left Finnbennach's shoulderblade there — from which comes Finniethe, the White One's Shoulderblade, as the name of that district. He drank again at Ath Luain, and left Finnbennach's loins there — that is how the place was named Ath Luain, the Ford of the Loins. He uttered a bellow at Iraird Cuillenn that was heard through the whole province. He drank again at Tromma, where Finnbennach's liver fell from his two horns — from which comes the name Tromma, or liver. He came to Etan Tairb and set his brow against the hill at Ath Da Ferta — from which comes the name Etan Tairb, the Bull's Brow, in Murtheimne Plain. Then he went by the Midluachair road to Cuib, where he had dwelt with the milkless cow of Dáire, and he tore up the ground there — from which comes the name Gort mBúraig, the Field of the Trench. Then he went on until he fell dead between Ulster and Uí Echach at Druim Tairb. So Druim Tairb, the Ridge of the Bull, is the name of that place.

Ailill and Medb made peace with Ulster and Cúchulainn. For seven years afterward none of their people was killed in Ireland. Finnabair stayed with Cúchulainn, the Connachtmen went back to their own country, and the men of Ulster went back to Emain Macha full of their great triumph.

FINIT. AMEN

NOTES ON THE TEXT

The notes that follow are designed to explain certain procedures and to give the minimum of additional information that seems necessary for an appreciation of the story.

BEFORE THE TAIN
page

1 HOW THE TAIN BO CUAILNGE WAS FOUND AGAIN. *Dofallsigud Tána Bó Cuailnge:* A ninth century anecdote, from the text in the Book of Leinster edited by H. Zimmer in the *Zeitschrift für Vergleichende sprachforschung . . . ;* Berlin 1887. Vol. XXVIII, pages 433/4.

Senchán Torpéist: Teacher of poets, said to have lived in the seventh century A.D.

Táin Bó Cuailnge: 'The Cattle-Raid of Cuailnge.' Cuailnge, anglicised Cooley, is the district south of Carlingford Lough in County Louth.

Letha: Latium — but here probably Brittany, which Irish monks and scholars traversed on their pilgrimage in Europe.

Cuilmenn: or *Culmen.* The *Etymologiae* of Isidore of Seville, a book revered at the time as the height of wisdom.

mac Roich: 'son of Roech.' No set name-form is followed in the translation, though the 'mac' is generally left untranslated in direct speech or where the relationship has no particular importance to the story. For consistency in the use of personal and place names the primary version given in Thurneysen's lists (appended to *Die irische Helden- und Königsage*) has usually been adopted.

If this your royal rock . . . : A passage of the *rosc* or *retoiric* verse discussed in the Introduction.

2 *Ulster:* 'Ulad' in the text. The modern anglicised form of the province-names is used in the translation. With other proper names, the Irish form is retained.

However, there are some who say . . . : A long and rambling account is given in a later story, 'Tromdámh Guaire', edited by Maud Joynt in vol. II of the *Mediaeval and Modern Irish Series*, Dublin Institute for Advanced Studies.

There are seven tales that prepare . . . : A different list is given in the text. The stories used in this translation have been kept to a minimum necessary for an understanding of the plot and motivations of the *Táin.*

3 HOW CONCHOBOR WAS BEGOTTEN, AND HOW HE TOOK THE KINGSHIP OF ULSTER. *Compert Conchoboir:* An eighth

century tale from the text in the manuscript Rawlinson B. 512, edited and translated by Kuno Meyer in *Hibernica Minora*, Oxford 1894, page 50; with extracts from the early twelfth century text *Scéla Conchoboir maic Nessa* in the Book of Leinster, edited by Whitley Stokes in *Eriu*, vol. IV, pages 22/33 — paragraphs 5 to 12, 15 (except the first two lines) to 17, and 21. A later and fuller version of *Compert Conchoboir* exists (edited and translated by Kuno Meyer in *Revue Celtique*, vol. VI) in which the mother is impregnated by swallowing a little creature in her drink, as in the stories, 'How Cúchulainn was Begotten' and 'The Quarrel of the Two Pig-Keepers'. The short version here has been chosen partly on grounds of quality and partly because, in the *Táin Bó Cuailnge*, there is not the same suggestion of the supernatural about Conchobor as there is about Cúchulainn and the two bulls.

...Eochaid Sálbuide of the yellow heel...: 'Sálbuide' means 'yellow heel'. An attempt is made throughout the translation to give the English meaning with these epithets where this is possible, and where the practice does not interfere too much with the narrative; similarly with place-names.

And at the feast of Othar she was delivered: Nothing is known about the feast of Othar. This short tale ends here, with the words 'and so on, as it follows in [the lost book] "The Iron Hauberk." '
  In a later version of the story Cathbad tells the mother at the last moment that if the child is not born until nightfall, its birth will coincide with the birth of Christ in the east. She manages to delay her son's birth by sitting on a stone slab.

Conchobor: There was a brook of this name in Crích Rois.

The boy Conchobor was reared...: This sentence is supplied from the later version (in *Revue Celtique* vol. VI) as a link passage with what follows.

Fergus mac Roich: 'Fergus mac Rossa' in the text. Rus was Fergus's grandfather.

4 *...and his foster-parents...:* It was normal for families of consequence to put their children out to foster-parents. (It will be noted that Conchobor does not recognise Cúchulainn, his own nephew, when he first appears at Emain Macha). Foster-parents were not necessarily of an inferior status; in the story 'How Cúchulainn was Begotten' the honour of rearing Cúchulainn is sought by the noblest among the Ulster heroes.

province: Literally, 'a fifth'; from an early division of Ireland into five fifths.

5 ... *the Twinkling Hoard:* Literally, 'speckled'.

6 *Gerg's vat:* The story is told in *Tochmarc Ferbe*, 'The Courtship of Ferb', edited with a translation by Ernst Windisch in *Irische Texte*, vol. III. There is an English translation by A. H. Leahy in vol. I of the Irish Saga Library, London 1902.

THE PANGS OF ULSTER. *Noinden Ulad:* From the text edited by Windisch in *Berichte über die Verhandlungen der Königlich Sächsichen Gesellschaft der Wissenschaften zu Leipzig: Philologisch-Historische Classe,* Vol. XXXVI, 1884. ('Uber die irische Sage Noinden Ulad').

Pangs: 'Noinden' (novena) is a period of nine days or other units. The phrase 'ces noinden' ('the nine days' affliction') is used in the *Táin* — where, it will be noted, the period of affliction lasts from autumn until the beginning of spring. It is generally held by scholars that there is some connection here with the practice of couvade, but it has also been suggested that there is a vegetation ritual involved, representing winter death and spring rebirth. In this translation the term 'pangs' is used throughout.

It is soon told: Literally, 'It is not hard to tell'.

8 EXILE OF THE SONS OF UISLIU. *Longes mac n-Uislenn:* An eighth or ninth century tale, from the text *Longes mac n-Uislenn,* edited by Vernam Hull: New York, London, 1949.

Uisliu: This is the version of the name most frequently used in the the text, but 'Uisnech' is also used.

9 ... *started up, ready to kill:* Literally, 'started away from each other in the house, face to face.' Changes of this sort, generally slight, have seemed necessary in order to produce a readable narrative. It is not proposed to note them all.

Parthian-red: 'partaing', a word of doubtful meaning. It has been suggested that it derives from 'Parthica'— Parthian leather dyed scarlet.

10 *son of Roech:* Fergus.

11 *Leborcham ... who couldn't be prevented:* Through fear that her verses might bring harm, Leborcham, as a satirist, would have more than usual freedom.

12 *'You will do it,' she said, binding him:* The words 'binding him' are not in the text. Her words put Noisiu under bond, or *geasa,* to do what she asked.

13 *Alba:* This means Britain generally, not only Scotland. It may also, as *Alpai,* be taken to mean the Alps, and so anywhere 'overseas'.

14 *So they were brought back to Ireland:* This sentence is added for
    the sake of clarity, together with a few other phrases later that
    bring out the nature of the compulsions — the *geasa* — governing
    the action. The original is so spare at this point as to be obscure.

17 *Indel:* the mother of the three sons.

20 *...and she was dead:* The following final sentence is omitted:
    'That is the exile of the sons of Uisliu and the exile of Fergus and
    the death of the sons of Uisliu and Derdriu. Finit.'

21 HOW CUCHULAINN WAS BEGOTTEN. *Compert ConCulainn:* An
    ancient — probably eighth century — tale, from the text edited by
    A. G. Van Hamel in *Compert ConCulainn and other stories:*
    vol. III of the *Mediaeval and Modern Irish Series*, Dublin Institute
    for Advanced Studies, 1956. His version I, from *Lebor na hUidre*,
    is taken, with the ending supplied from version II.

    Deichtine: In version I Deichtine is Conchobor's daughter. In
    version II she is his sister, and this is adopted for consistency with,
    e.g., Cúchulainn's identification of himself to Conchobor in chapter
    IV of the *Táin*.

22 *Brug on the Boann river:* The text gives only 'Brug'. This is the
    famous *Brug na Bóinne*, the dwelling by the river Boyne associated
    with the river goddess Boann and with the megalithic tomb at
    Newgrange.

23 *Lug mac Ethnenn:* A prince of the *síde*, a divine race said to have
    inhabited Ireland in earlier times. (They are associated later with
    the fairies.) The word *síde* applies primarily to the mound-dwellings
    in which they were believed to live.

    ...and called him Sétanta: It will be noted that anomalies of time-
    scale are ignored in the interests of giving Cúchulainn a triple
    birth.

    The men of Ulster were assembled...: The addition from version
    II begins here, replacing the ending of version I. Six lines of verse
    occuring a little later, and referring to the early part of the story
    as given in version II, are omitted.

24 *Blai Briuga, a landed man:* or 'hostel keeper'. He had charge of
    one of Ireland's great houses of hospitality. A *briugu* owned large
    herds and many ploughs, and enjoyed almost the same dignity as
    a noble.

25 *...until we reach Emain...:* It is stated earlier that they are in
    Emain already. It is not proposed to note all such inconsistencies.

CUCHULAINN'S COURTSHIP OF EMER, AND HIS TRAINING IN ARMS. *Tochmarc Emire:* Abridged from a tenth or eleventh century tale, from the text edited by A. G. Van Hamel in *Compert Con Culainn and other stories.* The full text would be disproportionately long among these preparatory tales and the following paragraphs have been omitted without, it is felt, spoiling the structure of the story: 1 to 3, 6 (in part), 10 (in part), 11 to 16, 17 (in part), 18 to 26, 27 (in part), 28 to 55, 58 (in part), 60 (in part), 61 to 66, 67 (in part), 71 (in part), 80 to 82, 83 (in part), 84, 85 (in part), 91 and 92.

Ol nguala: 'Iarngualae' (Ironcoal) is given in the text. The change is made for consistency with the end of the story 'How Conchobor was Begotten . . .', on page 6.

27 *May your road be blessed!:* Literally, 'I drive around you [in a chariot], turning to the right', i.e. in the direction of the sun's movement, and therefore attracting good fortune — compare the episode on page 60. To turn the left side of the chariot was an act of insult or challenge.

Then they spoke together in riddles: This sentence is in substitution for a long and detailed conversation, full of obscurities and 'kennings', in paragraphs 17 to 27 of the text.

. . . where the frothy Brea makes Fedelm leap . . .; who hasn't gone sleepless . . . sorrowing Autumn: Obscure. The translation of these and similar passages is very conjectural.

Samain . . . Imbolc . . . Beltine . . . (Lugnasa): The first days of November, February, May and August, when feasts introduced the seasons.

He finished his journey . . . : Omitted here is a long conversation between Cúchulainn and his charioteer explaining the mysteries of the earlier conversation (paragraphs 17 to 27) between Cúchulainn and Emer.

28 *the warped one from Emain Macha:* Cúchulainn. The reason for the nickname and the nature of Cúchulainn's 'warp-spasm' become clear in the *Táin.*

Gaulish: The word 'gall' later came to mean 'Norman,' 'the stranger,' but in the context of the Ulster cycle 'Gaul' seems the likely meaning.

Next morning the hero rose up . . . : 'Heroes' in the text. Laegaire, Conchobor and, 'some say', Conall Cernach, set out with Cúchulainn. They play little part in the story and are omitted altogether in the translation.

34 *the apple-feat . . . etc.*: The exact nature of these feats is not clear.
A few of the commentators' more reasonable guesses are embodied
in the translation.

the gae bolga: A terrible javelin used by Cúchulainn. Its use is
described in the Book of Leinster version of Cúchulainn's combat
with Ferdia:

> The *gae bolga* had to be made ready for use on a stream and
> cast from the fork of the toes. It entered a man's body with a
> single wound, like a javelin, then opened into thirty barbs.
> Only by cutting away the flesh could it be taken from that
> man's body.

imbas forasnai: Some form of divination or supernatural enlighten-
ment. One interpretation of the word 'imbas' suggests that the
method involved the use of the palms, but another simply means
'great knowledge.'

'*I salute you* — . . .': It is stressed that the translation of this
passage of *rosc*, and the others throughout the book, is very
conjectural.

37 *Cúchulainn returned to Emain* . . . : This and the next two sentences
are taken from a short episode concerning the rescue and betrothal
of Derbforgaill, daughter of Ruad, which intervenes at this point
(paragraphs 80 to 84 of the text) and is omitted.

39 THE DEATH OF AIFE'S ONE SON. *Aided Oenfir Aife*: A ninth or
tenth century tale, from the text in the Yellow Book of Lecan,
edited by A. G. Van Hamel in *Compert ConCulainn and other
stories*.

. . . *after Cúchulainn left Aife* . . . : This phrase is supplied in substitution for paragraph 1 of the text — a resumé of Cúchulainn's
relationship with Aife.

46 THE QUARREL OF THE TWO PIG-KEEPERS. *De chophur in dá
muccida*: A ninth century tale, from the text in the Book of
Leinster, edited by Windisch in *Irische Texte*, III Serie, vol. I,
pages 243 to 247. The order of sentences in the first paragraph is
changed in translation. The word 'cophur' in the title is obscure;
it has been taken as 'begetting' as well as 'quarrelling'.

síd: Singular of *síde*. See the note to *Lug Mac Ethnenn*, page 23.

Mongán mac Fiachna: 'The Shape-changer'; a re-birth of the sea-god
Manannán, and the subject of a group of stories. These are translated by Kuno Meyer in *The Voyage of Bran*, London 1897.

49 *Rucht and Friuch were their names* . . . : The text gives 'Rucne'
or 'Runce' here, instead of 'Friuch.'

Donn Cuailnge: This bull is called either Dub, 'the Black', or Donn, 'the Brown'.

50 *a hero to his herd at morning/foolhardy at the herd's head:* Literally, 'hero of the early herd/fool of the wandering herd.'

overlooks the ox of the earth: Literally, 'more than a hill of the ox of the earth.' Obscure.

THE TAIN

The chapter divisions and titles are the translator's.

52 I THE PILLOW TALK: This chapter is supplied from the twelfth century version in the Book of Leinster. The full text of the Book of Leinster *Táin* is available, with translation and notes, in *Táin Bó Cúalnge from the Book of Leinster,* by Cecile O'Rahilly; Dublin Institute for Advanced Studies, 1967.

Finn, the son of Finnoman, the son of Finnen . . . : There is no complete agreement on the form of these names.

53 *. . . and for every paid soldier I had ten more . . .* etc.: The 'nine' and 'four' are not in the Book of Leinster text, but they occur in another version.

Temair: Present-day Tara.

Conchobor . . . son of Fachtna: This is a reference to the later version of *Compert Conchoboir.*

light gold: 'findruine'. An alloy; white bronze.

58 II THE TAIN BO CUAILNGE BEGINS: The *Lebor na hUidre* Táin begins here. The text, with the Yellow Book of Lecan text, is edited by John Strachan and J. G. O'Keeffe, Dublin, 1912, and translated, with certain omissions, by Winifred Faraday in *The Cattle-Raid of Cualnge,* London, 1904. The line references throughout these notes are to the Strachan/O'Keeffe edition.

. . . to his brothers, the rest of Mágach's seven sons . . . : The text reads '. . . to Mágach's seven sons, i.e. Ailill, Anluan, . . . etc.'

a troop of three thousand: 'trícha cét'; literally, 'a thirty-hundred'; or it might mean the force of soldiers that a population of three thousand could raise. 'Trícha cét' is a territorial division.

Connlongas: Literally 'leader of the exile'— see 'Exile of the Sons of Uisliu'. Cormac's troop is erroneously given as three hundred men here in the text.

60 *. . . to the right, with the sun . . :* See note on page 27.

weaving-rod: A rod for weaving fringes. A magical weaving, connected with the art of prophecy, is implied.

61 *verse and vision:* 'filidecht'. The word means simply 'poetry' in modern Irish, but the functions of the ancient poet would have included prophecy.

62 *His jaws are settled in a snarl:* Conjectural. In the Book of Leinster version it is the points of his spears, not his jaws, that are bared. Compare the exchanges between Condere and Connla in 'The Death of Aife's One Son', pages 40/41

63 *the warriors/of Deda mac Sin:* The Clanna Dedad, or Degad; inhabitants of West Munster, with their capital at Temair Luachra. They were, under Cúroi mac Dáiri, one of the three heroic tribes of Ireland, the other two being the Gamanrad of Irrus Domnann and the Clann Rudraige of Emain Macha.

This is the way they went . . . : The descriptive phrases in this itinerary are translations of the names themselves ('by Tuaim Móna, the peat ridge'; 'through Finnglassa Assail, of the clear streams') or else attempts to bring out points of significance ('and across the Sinann river'; 'through Slechta, where they hewed their way'—'Slechta' deriving from the verb 'to hew or cut down'.)

The journey of the armies described in the text differs considerably from this itinerary.

65 *These are the places they were to pass . . . :* The order of these two sentences has been reversed. The following sentence is omitted: 'This is the end of the title-matter; the story in order now begins.'

III THE ARMY ENCOUNTERS CUCHULAINN.

. . . at Carrcin Lake: The lake is mentioned in a gloss. It is identified with the present townland and lake of Ardakillan, two miles south of Cruachan.

. . . their daughter, Finnabair: The name Finnabair is cognate with 'Guinevere'. It should not be confused with the place-name Finnabair in Cuailnge.

66 *Galeóin:* The explanatory phrase 'from north Leinster' is not in the text. The separateness of the Galeóin from the other 'men of Ireland' is not without significance. They belonged to a more ancient race, being said — like the Fir Domnann, to whom Ferdia belonged — to have come to Ireland with the Fir Bolg (*fir* = men) in the fourth of the five primeval invasions of Ireland. It has been suggested that there is a connection between the Galeóin and the Gauls (as between the Fir Bolg and the Belgae, and the Fir Domnann and the Dumnonii). The matter is discussed at length in T. F. O'Rahilly's *Early Irish History and Mythology*.

68 *until that warrior's work is done:* The Yellow Book of Lecan text of the *Táin* begins here.

Nemain: 'the war-spirit' is not in the text. Nemain (Panic) is one of the three goddesses of war, the others being Badb (Scald-crow), haunter of battle-fields, and Morrígan (Great Queen or Queen of Demons). Nemain and the Badb are mentioned as wives of the war-god Nét.

You must go and warn Ulster: It will be noted that Sualdam vanishes from the story at this point. He reappears, with his warning, only toward the end of the story, in chapter XII.

a spancel-hoop of challenge: A ring made from a twig and used for spancelling horses and cattle. Such hoops were left at fords or boundaries as messages of challenge.

ogam: An incised lettering used on stone or wood.

71 *The druids answered:* This phrase is not in the text.

72 *Partraigi:* A tribe said to have been related to the Dumnonii.

73 *(It is from this . . . forked branch):* This occurs three sentences later in the text.

. . . waiting for them at the ford: A ford is frequently the place of challenge and single combat. In a practical sense it would be natural to defend a boundary, following a river, at such a crossing-place, as in this episode; but warriors appear also to select a ford in a more formal way, as an arena for certain kinds of combat — as in the combat between Cúchulainn and Fraech in chapter V, the combat with Lóch in chapter VIII, with Ferdia in chapter XI, and so on. Looking at the symbolism of the matter, and taking account of the mysterious nature of boundaries in themselves, Alwyn and Brinley Rees, in *Celtic Heritage* (page 94), suggest that the ford combat 'partakes in some measure of the nature of a divination rite.'

75 *— none like Cúchulainn:* The following sentence in the text is omitted: 'These are the Praises of Cúchulainn.'

IV CUCHULAINN'S BOYHOOD DEEDS.

. . . in their oaken house . . . : Conjectural.

77 *fidchell:* Some kind of board game.

79 *the Badb:* See note to *Nemain*, page 68.

88 *the del chliss:* Literally: 'the dart of feats'.

91 *. . . when he is fully seventeen years old?:* The following sentence is omitted: 'These above are the boyhood deeds of Cúchulainn on the Táin Bó Cuailnge.'

V 'DEATH DEATH!': The following introductory sentence is
omitted: 'A separate version follows, to the death of Orlám.'

92 *Ath Froich is the name of that ford still:* In the text this occurs
three sentences later.

   ...a troop of women in green tunics..: Fraech's mother Béfinn
   was a woman of the *síde*, sister of the Boyne river-goddess Boann.

93 *Druim Baiscne:* 'Baiscne' is not in the text. Omitted here is the
following sentence referring to later events:
   But according to another version it is there that the squirrel
   (or weasel) that Medb had in her chariot, and her pet bird,
   were killed with sling-stones; in the present version this
   happens after the killing of Orlám.

94/5 *...thinking it was an Ulster warrior:* This is added for clarity.
The sentence is slightly recast in translation.

96 *...he'd break my head with a stone:* Omitted here is a scribal
note about 'a separate passage' which gives the fight against
Gárach's three sons and the end of the episode of Orlám's charioteer. The order of these items has been reversed in translation.

   ...outside the camp: Not in the text. These words are supplied to
   try to clarify Cúchulainn's motive in killing the charioteer.

   Reuin: This person seems to owe his brief existence to a mis-
   reading of 'nEuin' (a few lines earlier). It is not the only invention
   of the kind — see the note to 'Lethan' following.

97 *Lethan ... Ath Lethan:* Lethan is apparently an invention. 'Lethan's
Ford' would read 'Ath Lethain'; 'Ath Lethan' simply means
'the broad ford.' Haley believes the 'broad ford' crossed an inlet to
the north of Dundalk Harbour.

   ...in a rage at what Cúchulainn had done...: Presumably the
   killing of his foster-son Meslethan and the five others at Ath meic
   Gárach.

   ...the previous ford...: Literally: 'the ford near it.'

98 *the Morrígan:* See note to *Nemain*, page 68.

100 *...tossing the earth back over him with his heels:* Supplied from
the Book of Leinster version.

VI FROM FINNABAIR CHUAILNGE TO CONAILLE.

100/1 *It is said in one version of the tale ... It is further said, in this
version ... etc ...:* These phrases are added for clarity. In the text
the 'first version' is given without comment until the army has
reached Druim Féne in Conaille (page 102), at which point the

second version of the journey, written in a more elaborate style, begins.

...the armies divided...: See the ending of chapter II.

...Lóthar, was summoned: The following sentence is omitted here: 'The following is from a separate version.' The next sentence in the translation is from the version in *Lebor na hUidre* only. In the Yellow Book of Lecan two earlier sentences are repeated here; they are omitted in the translation.

101 *Medb asked the herdsmen...*: 'herdsman' in the text.

...so that the water wouldn't force him backward... and he drowned: These two phrases are supplied for clarity from a later version of the episode in a fragmentary manuscript H.2.17 in Trinity College, Dublin.

It is there that Cúchulainn killed Cronn and Caemdele... at the same river: This passage and the paragraph which follows it in the translation occur in the reverse order in the text.

Some say that this is the reason... long afterward: This sentence is supplied from H.2.17.

105 *buanbach*: Like *fidchell*, some kind of board game.

111 *horsemen*: *marcach*, 'horseman'. There are scholars who are not convinced that warriors at this time went on horseback.

112 *...I'd give you one and share another*: Literally, 'you shall have one and the half of another.' The meaning of this formula is not clear.

113 *'Lugaid,' Fergus said, 'will you do this for me...'*: There is a slight confusion in the text as to the speaker of the previous sentence. The words 'Fergus said' are inserted here for clarity.

...toward Cúil Airthir, in the east: A passage of thirteen lines is omitted after this, relating how Cúchulainn went that night to report to Conchobor that women are being captured, cattle driven away and men slain by Ailill and Fergus; he returns to see the army setting off. In this version, the passage is an anticipation of Sualdam's warning to the men of Ulster (page 219).

114 *Cuillius...Cúchulainn struck him with a stone and killed him*: Cuillius reappears late in the tale, when Fergus recovers his sword (page 246).

VII SINGLE COMBAT.

116 *A heavy snow fell that night...*: This sentence is supplied from the book of Leinster version.

A dark, good-looking, broad-faced man ... the marks of a herald:
Supplied from the Book of Leinster version.

117 *... after each combat. This plan will gain time for him ... :* These
words are supplied for clarity.

121 *Cúchulainn murdered no more that night with his sling:* This is
followed in *Lebor na hUidre* by the sentence 'and the women
and maidens and half the cattle were sent to him, and his food
was sent daily.' This is omitted as inconsistent with the earlier
arguments leading to the plan of single combat. It is probable that
another version of the story is involved here. Compare the episode
in note (a) to the long interpolation beginning on page 127.

122 *'We'll see,' Cúchulainn said:* The following sentence is omitted: 'It
was here that Cúchulainn gave the chant beginning: "If I survive
Nadcranntail etc.".'

Cúchulainn was there before him in the distance ... : The words
'in the distance' are added to account (to some extent) for Nadcranntail's doubt about Cúchulainn's identity next day.

123 *... with his weapons in their wagon:* From here until Nadcranntail
goes to speak with his sons (page 124), Thurneysen believes two
versions of this episode are fused. In this translation the text is
accepted as it stands, but the sentence 'and it struck the standing
stone and shattered in two' is brought forward from the second
encounter to the first, after 'Cúchulainn leaped on high', in an
effort to reconcile the inconsistencies.

125 *... a third of Medb's men:* The following is omitted here: 'Such
was the death of Nadcranntail.'

VIII THE BULL IS FOUND. FURTHER SINGLE COMBATS. CU-
CHULAINN AND THE MORRIGAN.

the Picts: The Cruithni, Pictish tribes of Down and Antrim.

coming from the direction of Sliab Cuilinn: In the text: '... Buide
mac Báin (literally, 'Yellow son of White') from Sliab Cuilinn with
the bull and fifteen heifers about him and three score ... etc.' The
slight alteration is made for clarity; it is told in chapter VI how
the bull escaped to the fastnesses of Sliab Cuilinn.

126 *... as you shall hear in the proper place ... :* This phrase is not
in the text. The incident referred to occurs on p. 138.

127 *... who had been captured with the bull ... :* This phrase is added
for clarity, referring to Buide mac Báin's statement on p. 125.

... the herd after him ... : 'him' in the text.

...if only a man could be found to withstand Cúchulainn at the ford: Lines 1362 to 1486 of the *Lebor na hUidre* version are omitted here. The passage, which does not occur in the Yellow Book of Lecan, is interpolated with little regard to consistency or sequence of events. It is made up of a number of episodes and is dealt with in this translation as follows:

(a) lines 1362 to 1379: omitted entirely. The episode reads:

'Ask Cúchulainn for a truce,' Ailill said. 'Send Lugaid to him,' they all said. So Lugaid went to speak with him. 'How do I stand now with the host?' Cúchulainn said. 'They think your demands are a great insult,' Lugaid said, 'the women and girls and half the cattle. And hardest of all is being killed while keeping you in food.' A man fell there each day until a week was passed. They played Cúchulainn foul, sending twenty to fall on him all together, but he killed them all. 'Go and ask him for a change of place, Fergus,' Ailill said. Then they moved to Cronech. The following fell in single combat there: two named Roth, two named Luan, two thieving women, ten jesters, ten cupbearers, ten named Fergus, six named Fedelm, six named Fiachra. He slew all of these in single combat. When they first pitched their tents in Cronech they discussed what to do with Cúchulainn. 'The best thing to do,' Medb said, 'is this: let me send someone to ask him for a truce with our host, and offer him half of our cattle.' This message was brought to him. 'I agree,' Cúchulainn said, 'for as long as you don't break the pact.'

(b) lines 1380 to 1414: Cúchulainn's meeting with Finnabair, and the death of Tamun. This passage is given in chapter IX, pages 140/1. ['Take this message to him . . .' There was no further truce for them with Cúchulainn after that.] Lines 1381 to 1395 are omitted, however: these contain a sketchy repetition of the incident of Mac Roth's first approach to Cúchulainn (see chapter VII, page 116), with Maine Athramail replacing Mac Roth. The passage reads as follows:

Maine Athramail, the Fatherlike, went to him and met Laeg first. 'Whose man are you?' he said. Laeg didn't speak. Maine spoke to him again three times. 'I am Cúchulainn's man,' he said at length, 'and if you bother me I will take off your head.' 'A fierce man!' Maine said turning from him. Then he went to speak with Cúchulainn. He found him stripped of his shirt, squatting in the snow waist-deep; and the snow was melted about him with the hero-heat for a man's span on all sides. Maine asked him too, three times, whose man he

was. 'Conchobor's man; and don't annoy me. If you annoy
me further, I'll take off your head like a blackbird's.' 'It isn't
easy talking with this pair,' Maine said. He left them then and
told his story to Ailill and Medb. 'Send Lugaid to him,' Ailill
said, 'to offer him the girl.'

(c) lines 1415 to 1428: the combat of Munremur and Cúroi. This
passage is given in chapter XII, pages 216/8. See note to those pages.

(d) lines 1429 to 1433: omitted entirely; it reads:
    'Tell Cúchulainn,' Medb and Ailill said, 'that we want to move
    to another place.' He agreed to that, and they changed
    position. Then the Ulstermens' pangs left them. Immediately
    they had started up from their pangs some of them set out
    against the armies to join in the slaughter again.

(e) lines 1434 to 1450: a version of the death of the boy-troop.
The material is used later, fused with another version, in chapter
IX. See notes to page 144.

(f) lines 1450 to 1456: a brief description of Cúchulainn's warp-
spasm; omitted entirely.

(g) lines 1457 to 1479: a version of 'Rochad's Bloodless Fight.'
See note to pages 214/5.

(h) lines 1480 to 1486: the death of the princes. The beginning of
chapter IX; see note to page 137.

With this interpolation removed, the incidents that now follow
the death of Forgaimen in this chapter are the same, and in the
same order, as in the Book of Leinster.

128 ... *through the back of the fierce warrior's head:* The following
sentence is omitted here: 'Or it was in Imslige Glendamnach that
Cúr fell, according to another version.'

130 *Ferbaeth, your fool's foray ...:* 'Ferbaeth' means literally 'foolish
man.'

131 *'I wonder will you pay ...' 'Sometime I must pay':* The words 'I
wonder' and 'sometime' are not in the text. They are added to give
some clearer sense; Thurneysen thinks the idea of 'Wergeld' may
be involved.

And Cúchulainn went and talked with them: The phrase so translated seems, in the text, to be part of Lugaid's speech. It is not
clear what is going on, unless there is fairly free movement in
and out of Medb's camp at this stage. See, for example, Fergus's
movements after the death of Cúr; Fergus is also in Cúchulainn's
presence after Ferbaeth's death.

132 *Cúchulainn beheld at this time a young woman ...:* The words
'at this time' are not in the text. Cúchulainn's meeting with the
Morrígan is the subject of a separate preparatory tale, *Táin Bó
Regamna*. It seems clear that the incident is interpolated in the
Táin here at the last moment; the denouement comes in the combat
with Lóch mac Mofemis which follows. As the version given here
is preferred to the *Táin Bó Regamna*, it has been decided to leave
the *Lebor na hUidre*/Yellow Book of Lecan arrangement as it stands.

133 *Then she left him:* The following sentence in *Lebor na hUidre* is
omitted here: 'He was a week at Ath Grencha, that is Ath Darteisc.'
A title for the following episode is also omitted.

  ... he would kill him for it: A short section in *Lebor na hUidre*
which does not occur in the Yellow Book of Lecan, is omitted
here. It anticipates the episode on pages 158/9, ['This is a nuisance,'
she said ... Then Medb herself climbed on the men's backs to see
him.] and begins:
  'Have done with fair play,' Medb said to her followers. 'Go
  against him in force eastward on the ford.' The first warriors
  to go were the seven Maines, and they saw him on the western bank of the ford. He was wrapped in festive raiment, and
  the women climbed on the men's backs to see him.

134 *He did this, to get Lóch to fight him:* A section of 65 lines
follows in the *Lebor na hUidre* version; it does not occur in the
Yellow Book of Lecan. The component parts are dealt with in
this translation as follows:
  (a) lines 1643 to 1655 are retained here despite a certain slight
  confusion [And he plucked a fistful of grass ... his week's deeds
  at Ath Grencha.]
  (b) lines 1656 to 1702 —'the fourteen at Focherd'— are given in
  this translation after the death of the princes at the beginning of
  chapter IX (pages 137/9) [Medb considered again what to do ...
  and four named Dúngas from Imlech]
  (c) lines 1703 to 1708 are retained here [Then Medb began to
  incite Lóch ... you both learned your skill.]

  ... and spoke into it ...: The original suggests he is casting a spell.

135 *Tarteisc:* A typically tortuous derivation. An alternative meaning,
suggested by Thurneysen, is 'water-thirst.'

135/6 *So it was that Cúchulainn did to the Morrígan ...:* This sentence
occurs after Cúchulainn's chant in the text; the words 'on the
Táin Bó Regamna' are omitted.

136 *Then Cúchulainn cut his head off:* Eight lines are omitted here

and transferred — with a further interpolation, as will be noted — to the next chapter (pages 139/40), after the 'fourteen at Focherd' [Afterward, on the same day ... between Delga and the sea].

137 *The blessing of god and man ...:* Literally, 'of gods and non- (or lesser-) gods.' The following scribal comment is omitted:

> Their 'gods' were powerful beings, while their 'lesser-gods' watched over the tilling of the soil. But others say that he gave her his blessing-judgment, saying over every item 'A blessing-judgment upon you.'

'...I wouldn't have done it,' Cúchulainn said: The following sentence is omitted here: 'This may be the story called "Riamdrong ConCulainn" (Cúchulainn's Advance Troop) at Tarteisc on the Táin Bó Cuailnge.'

IX THE PACT IS BROKEN: THE GREAT CARNAGE: The first part of this chapter is made up of sections transferred from other parts of the text. They are:

— the death of the princes ['Ask Cúchulainn for a truce'... and Cúchulainn slew all six.], episode (h) of the first interpolation removed from chapter VIII (see notes to page 127);

— the fourteen at Focherd [Medb considered again ... and four named Dúngas from Imlech.], episode (b) of the second interpolation removed from chapter VIII (see notes to page 134);

— lines 1739 to 1746 omitted from the previous chapter after the death of Lóch [Afterward on the same day ... between Delga and the sea.]; see notes to page 136. Inserted in this passage in the translation are lines 1759 to 1772 [Fergus said they must stop ... and the Ford of Cét Chuile.], which seem more in place in the events leading up to the arrival at Delga;

— Cúchulainn's meeting with Finnabair, and the death of Tamun ['Take this message to him ... There was no further truce for them with Cúchulainn after that.]; episode (b) of the first interpolation removed from chapter VIII.

At this point, as the four provinces of Ireland settle down on Murtheimne Plain, the translation returns to the order of events in the text.

138 *...fourteen of her own most skillful followers:* The names that follow come to twenty unless the 'two sons of Buccride' are the 'two named Glas Sinna', and similarly with the next two pairs. In episode (a) of the first interpolation omitted from chapter VIII there is a reference to an attack on Cúchulainn by twenty men followed by the move to Cronech (Focherd). In the Book of

Leinster version the list of names comes to a total of seventeen, and these are called 'the men at Cronech ... the same as the twenty men at Focherd'. There is some confusion about the whole episode of the breaking of the pact of single combat.

139 ... *and Cúchulainn killed them single-handed:* The following sentence is omitted here:

> It may be that 'the Fortnight [or Fourteen] at Focherd' and 'the Five in One Field' refer to these things, or it may be that 'the Fortnight at Focherd' on the *Táin* comes from Cúchulainn's spending fifteen days in Focherd.

— *at that time called Dún Cinn Coros:* An obscure line is so deciphered by Thurneysen.

140 *three druids:* 'two' in the text.

... *the last Battle:* The final battle of the *Táin* at Gáirech and Irgairech (chapter XIV).

Tamun: The fool is not named. His name is given, however, in a later version of the episode (lines 2129 to 2132), omitted from this translation, which reads:

> Ailill's people placed the king's crown on Tamun the fool, Ailill being afraid to wear it himself. Cúchulainn flung a stone at him and it smashed his head. From this comes the name Ath Tamuin, Tamun's Ford, and Tuga in Tamuin, the Covering (or Disguising) of Tamun.

141 *The four provinces of Ireland settled down ...:* Here the narrative returns to the order of the text, which continues in the elaborate later style of the Book of Leinster version. From this point to page 163 the episodes, with certain omissions and substitutions, are more or less word for word as in the Book of Leinster.

142/4/8 *His feats and graceful displays ... Cúchulainn's aching wounds and several sores ... his warlike battle-weapons ...:* Literally, 'the play and sport and diversion ... the wounds and cuts and gashes and many injuries of Cúchulainn ... his weapons of battle and contest and strife ...'. The translation does not try to reproduce exactly the groups of more or less interchangeable nouns (or adjectives where they occur) which are a mannerism of the later version of the *Táin*.

143 ... *as men sing to men ...:* ferdord; literally, 'the man-murmur.'

144 *The boy-troop in Ulster spoke among themselves at this time ...:* Episode (e) of the first interpolation omitted from chapter VIII is fused here with the following version of the episode which occurs at this point:

Now it was that the boy-troop came down from Emain Macha
in the north, thrice fifty sons of Ulster kings, led by Follamain, Conchobor's son. They waged three battles on the armies
and slew three times their own number, but they themselves
fell, all but Follamain mac Conchoboir. Follamain swore he
would never, to the very edge of doom, go back to Emain
unless he took Ailill's head with him, with the gold crown on
top. In this he swore no easy thing, and the two sons of
Bethe mac Báin, sons of Ailill's foster-mother and foster-
father, went out and set upon him and he perished at their
hands. That is the death of the Ulster boy-troop and Follamain
mac Conchoboir.

Lines 1450 to 1456, giving a brief description of Cúchulainn's
warp-spasm, are omitted here, with a sentence identifying Lia Toll
as Lia Fiachrach mac Fir Febe, so named because Fiachra died
there.

150 *the baying of a watchdog at its feed:* The phrase translated 'at its
feed' is obscure. With these extravagant descriptions, as with the
rosc or other verse passages, greater liberties are taken elsewhere.

153 *In that style, then, he drove out to find his enemies...:* This
occurs in the text at the beginning of the next sentence.

156 X COMBAT WITH FERGUS AND OTHERS.

bards: 'aes dána. 'men of art.'

158 *seven bright pupils, eye-jewels:* 'pupils' is not in the text.

a gryphon's clench: The meaning of the word 'griúin' is not certain.
It can mean hedgehog or gryphon.

'This is a nuisance,' she said...: See note to page 133.

159 *tales to tell:* Obscure. The translation is conjectural.

nine heads in one hand: 'eight' in the text, but this is changed to
match the previous description. It is 'nine' in the Book of Leinster.

160 *would put an end/to this Warped One:* Supplied from the Book of
Leinster.

the maiden-massacre: See 'Exile of the Sons of Uisliu' (page 15).

lags/in the battle against Ulster: Literally, 'does not contend for
the kingship of Ulster.'

163 *Fiacha Fíaldána, the bold and true...:* In this curious incident
there appears for the moment to have been a changing of sides.
Dubthach is presumably wounded and not killed, and appears

later in the *Táin*. Here, and more clearly in the alternative version which follows — and in the episode of Aengus after that — there is the first indication that the men of Ulster are beginning to rise from their pangs.

...and have them fight, as Medb promised: There is no such promise in this version of the *Táin*. This whole group of episodes seems to come from a different source.

...when he came back: not in the text.

164 *— the Height where the Armies were:* Here, as already noted, (note to page 140), the second version of the death of the fool Tamun is omitted.

...at Muid Loga: The Connacht armies are now in full retreat.

...before they came under the sword at Emain Macha: There is no battle at Emain Macha in this version.

...with no sword in your scabbard: The following sentence is omitted here: 'For Ailill had stolen it, as we have told.'

165 *Ferchu Loingsech:* This is the Book of Leinster version. The Yellow Book of Lecan/*Lebor na hUidre* version reads:

> Then Ferchu Loingsech, 'the Exile', who was in exile because of Ailill, hears them. He goes against Cúchulainn, his troop numbers thirteen men. Ferchu is killed at Cingit Ferchon, 'Ferchu's Goblet'. Their thirteen standing-stones are there.

166 *...the place of Ferchu's Head:* The following episode is omitted here. It occurs only in *Lebor na hUidre*, and is listed by Thurneysen among the interpolations:

> Medb sent out Mann, son of Muresc mac Dáiri, of the men of Domnann, to fight Cúchulainn. He was a brother of Damán, Ferdia's father. Mann was a harsh man, gross in fighting and sleeping, and foul-tongued — he was as ill-spoken as Black Dubthach of Ulster. He was strong and stout, and as steady of limb as Munremur mac Gerrcinn; a strong fiery man like Triscoth, that strength of Conchobor's house.
>
> 'I'll go unarmed and crush him in my hands. Where is the honour and glory if I go to work with my weapons on this barefaced sprite?'
>
> He went out to find Cúchulainn. Cúchulainn and his charioteer were out on the plain keeping watch on the host.
>
> 'There is a man coming toward us all alone,' Laeg said to Cúchulainn.
>
> 'What kind of man?' Cúchulainn said.
>
> 'A dark, strong, black, bull-like man, unarmed.'

'Let him pass,' Cúchulainn said.

Then he came up to them.

'I have come for fight,' Mann said.

They took to wrestling then for a long while and Mann threw Cúchulainn three times. The charioteer scolded Cúchulainn:

'You'd be fierce enough with the boys in Emain if you were fighting there for your champion's share,' he said.

Heroic rage seized Cúchulainn and his battle ardour rose, and he hurled Mann at a standing stone and burst him asunder. Hence comes the name Mag Mannachta, the Plain of Mann's Death — reading it 'Mann echta,' the death of Mann. With this episode the *Lebor na hUidre* version ends. The story continues in the Yellow Book of Lecan.

166 ... *twenty-nine men:* 'twenty-seven' in the text, but the total in the episode comes to twenty-nine.

They were Gaile Dána and ...: The beginning of this sentence is omitted [They cast their twenty-nine javelins against him all together, viz. Gaile Dána ... etc.] The following sentence is also omitted: 'When they all stretched out their hands, Fiacha mac Fir Febe came after them out of the camp.' Part of the more explicit Book of Leinster version is substituted [The arrangement to do this ... except those that lay asleep — pages 166/7]. The conclusion is given from the Yellow Book of Lecan — whose version of names, etc., is retained throughout.

Every one of them has poison on him ...: It is through the children of Calatín (as the name is given in the Book of Leinster) that Cúchulainn, in the story *Aided ConCulainn*, or *Brislech mór Maige Muirthemne* (The Death of Cúchulainn, or the Great Carnage on Murtheimne Plain) meets his death.

167 ... *the compact is broken now for the Ulster exiles:* Fiacha, one of the exiles, having committed an act of war against Connacht warriors.

... *with the two sons of Ficce* ...: These appear again, as if for the first time, in the first episode of chapter XII (i.e. in the very next episode, allowing that the combat with Ferdia is an interpolation — see the following note). There seems to be some confusion; it is possible that in its place here the name 'Ficce' is an echo of the following passage in the Book of Leinster:

One of them, Glas mac Delga, took to his heels and escaped while Cúchulainn was beheading all the others. Cúchulainn gave chase. Glas ran ahead of him around Ailill and Medb's

tent, but he had only said 'Fiach! Fiach!' ('A debt! A debt!') —
intending to betray Fiacha — when Cúchulainn struck him and
took off his head. 'They made quick work of that one,' Medb
said. 'What debt was he talking about, Fergus?' Ailill said. 'I
don't know,' Fergus said, 'unless someone in the camp owed
him something and this was on his mind. Or perhaps a debt of
flesh and blood. Anyway,' Fergus said, 'he has been paid all
his debts together now.'

This passage has its own dubious features — the casual reference,
for example, to what would have been a momentous event,
Cúchulainn's penetration of the Connacht camp.

168 XI COMBAT OF FERDIA AND CUCHULAINN: This episode, the
last of the single combats, is an interpolation, of not earlier than
the eleventh century, into the tale. It is a compendium of elements
used in earlier episodes. Even more than the 'Boyhood Deeds',
chapter IV, it has the characteristics of a separate work. The Book
of Leinster version is used as noted below.

169 ... *offered him certain fragrant sweet apples* ... : Literally, 'dis-
tributed very fragrant apples to him across the bosom of her
shirt.' The ambiguity seems deliberate.

170 *Does it take your breath away:* In the original the meaning appears
to be: 'You will get more than your breath' (in the sense of
expectation).

... *who can ensure/whatever you ask:* A missing line supplied
from the Book of Leinster.

171 *Take farmers or soldiers* ... : This stanza and the next come in
reverse order in the text. The order adopted here is the same as
in the Book of Leinster. The line 'or Niaman the slaughterer' is
supplied from the Book of Leinster.

172 *queenly Finnabair:* Literally, 'Finnabair, queen of western Elga.'
(Elga was one of the ancient names of Ireland.)

... *and found Cúchulainn:* A passage of formal description is
omitted here. It is a repeat, in more elaborate terms, of the description of Fergus's arrival with Etarcomol in chapter VII (page
118). It has its place in the Ferdia episode considered as a separate
work, but not in the context of the *Táin* as a whole. (The sword
in Fergus's scabbard is described, for example, though his scabbard
should be empty at this point in the story.)

'*I believe I am,' Fergus said:* The following is omitted here for the
same reason as in the previous note:

'You are right to believe it,' Cúchulainn said. 'If a flock of

birds were to settle on the plain, I would give you one wild
goose and share another; if the fish were running in the
river-mouths, I would give you one and share another; with
a fistful each of cress and marshwort and sea-herb, and
afterward a drink of cold water out of the sand.' 'That is
outlaws' food,' Fergus said. 'So it is,' Cúchulainn said; 'and
mine is an outlaws' lot, with not one night of ease all the
time I have been here — from the Monday after Samain at
summer's end until today — but strongly withstanding the
men of Ireland now on the Táin Bó Cuailnge.' 'If it was food
we wanted,' Fergus said, 'we would have stayed where we
were. But that isn't why we came!'

In the next sentence of the translation the word 'else' is omitted
('What else do you want here?') because of the omission of this
passage.

175 *As to Ferdia, he went to his tent . . . :* The rest of this sentence is
added from the Book of Leinster; it gives some explanation of the
'surety' arrangement in the chant between Medb and Ferdia (pages
170/1).

179 *Who says he is not . . . :* Literally, 'it is known, let there be no
silence (about it).' This line is supplied from the Book of Leinster.

I hear him, and he hears: The following sentence is omitted here:
'Here is a description of Cúchulainn's chariot: the third main
chariot description on the Táin Bó Cuailnge.'

'How does Cúchulainn look?' Ferdia said to his charioteer: Here
as with the arrival of Fergus, a description of Cúchulainn, his
chariot and his charioteer is omitted (lines 2548 to 2577). There
is nothing in this passage that is not in the description leading
up to the Great Carnage, in chapter IX.

180 *. . . He hasn't paid you anything:* The sentence 'That is how he
gave the description' is omitted.

We'll dispose of him . . . : The meaning is roughly 'we will supply
what is needed.' This line does not occur in the Book of Leinster
version, which has instead 'for it is for the sake of reward.' The
latter line is in the Yellow Book of Lecan also, but is superfluous
and omitted from the translation.

180/1 *. . . and Ferdia said to Cúchulainn: 'You are welcome . . . ':* Here
the narrative in the Yellow Book of Lecan breaks down abruptly
and the episode is disposed of in a further 124 lines — 25 lines of
prose giving fragments of the action and 99 of verse, much of
which is in the Book of Leinster version. The latter is adopted

from this point on, except for the chant: 'Ferdia of the hosts ...'
— see note to page 201.

181 *Squinter: Cúa;* this could be a diminutive of *Cú*, 'Hound', but is
glossed in the Yellow Book of Lecan version as follows:

> *Cúa* was the word for squinting in old Gaelic. There were
> seven pupils in Cúchulainn's eye. Two of those pupils were
> squinting, which was no more a blemish in him than it was
> a good feature.

182 *'Death has seized you!':* Conjectural. Cecile O'Rahilly suggests: 'A
cancer has attacked them.'

185 *... you alone/could escape ...:* Conjectural. The literal meaning
appears to be 'why was my combat turned against you alone?'

186 *if I were the one that Medb smiled at:* possible readings seem to
be: 'upon whom the fair head of the province — or all the fair
province — smiled.'

... touch the least part of your flesh: Literally, 'redden your side,
north or south or east or west.'

190 *... to Cúchulainn on the northward side of the ford:* The rest of
the repetition is omitted ('for there were more supplying Ferdia ...
and talk with him.')

195 *... as if he was rinsing a cup in a tub ...:* Conjectural. Cecile
O'Rahilly suggests it is the beating of flax in a pond.

196 *... he called out to Laeg mac Riangabra for the* gae bolga: A brief
description of the *gae bolga* is given next in the text. This is
omitted here in the translation, but is included in the notes to
'Cúchulainn's Courtship of Emer ...' (page 34).

'Beware the gae bolga!' *he said:* This phrase is added from the
Yellow Book of Lecan version.

196/7 *the highways and byways of his body:* or the fundament.

197 *'I am leaving this life':* Conjectural. Cecile O'Rahilly suggests
'heroes have been destroyed.'

201 *Ferdia of the hosts ... etc.:* This chant is given in the Yellow Book
of Lecan version. ['of the hosts'— reading with the Book of
Leinster version.]

a sight to please a prince: reading with the Book of Leinster version.

... golden brooch, I mourn: Following this in the Yellow Book of
Lecan text there are eight lines of verse from Cúchulainn's final
lament, and nothing further. (Neither these, nor the next 238
lines of the Yellow Book of Lecan text — making the first twelve

pages of chapter XII of this translation — are translated by Winifred Faraday.)

204 *...the same rights, the same belongings...*: Conjectural. Cecile O'Rahilly suggests 'the same power of guarantee.'

205 *...Misery! A pillar of gold...*: reading with the Yellow Book of Lecan version.

Banba: one of the three eponymous goddesses — Eire, Banba and Fodla — whose names are associated with Ireland.

...a king's son of fairer fame: In the text the episode ends with the further phrase: 'The death of Ferdia, to this point.'

206 XII ULSTER RISES FROM ITS PANGS: The translation resumes here from the Yellow Book of Lecan (line 2734). It will be seen that there is still some confusion. This arises from the same break in the manuscript which affected the Ferdia narrative (see the note to pages 180/1 — and see also the note to page 207 below).

...the river Sas, for ease...etc.: The proper names only are given in the text.

...close by the Mash of Marrow, of which you shall hear: These words are not in the text. 'Imorach Smiromrach' means 'the edge of the Bath of Marrow.'

Mac Roth...went northward: It is 'southward' in the text.

207 *...that he saw only one chariot:* The following note appears on the manuscript at this point:
> The man who wrote that section of the Táin Bó Cuailnge [presumably the Ferdia episode] was a choice writer.
> My name: Cathal Marti ix° M.D.CC. LXX.

A list of titles of the final high points of the *Táin* occurs here, and is omitted. Titles of the individual episodes in this chapter are also omitted, unless as head notes.

208 *They had no right to give me bad news:* The following note is omitted:
> because every healer who examined him said he wouldn't live, that he was completely incurable, which is why he struck them with his fist.

They sent for the holy healer Fingin...etc.: This sentence occurs three sentences earlier in the text. The phrase translated 'the holy healer' means literally 'the seer and healer.'

209 *two forest kings...the two sons of the forest king:* or 'of the king of Caill'. This reference, and 'the two sons of three lights', is obscure.

. . . their faces the size of wooden bowls, one bigger than the other:
Obscure. 'Faces' is not in the text, but Cúchulainn's face is likened
to a bowl, in much the same terms, in his warp-spasm (page 150).

210 *Iruath:* Identified as Norway; some northern land.

211 *Cormac, Mael Foga's son:* This is the Book of Leinster reading; the
reading in the Yellow Book of Lecan, 'Cormac maile ogath', seems
corrupt.

gold crowns: reading with the Book of Leinster.

212 *Ochtur Lui, in Crích Rois . . . :* It will be noted that this puts
Cúchulainn, at the present stage of the story, in Crích Rois and
not Conaille (compare the first paragraph of the chapter.)
Smirommair, where the Cethern episode takes place, is identified
as Smarmore, a few miles south of Ardee (Ath Firdia).

213 *. . . his son Crimthann:* The name Crimthann is given in the Book
of Leinster version.

*When Fintan's people and the men of Ireland were found . . . The
men of Ireland said: 'It is a red shame . . .':* These sentences are
supplied from the Book of Leinster.

214/5 *Cúchulainn told his charioteer to go for help to Rochad . . . The
girl slept with him that night. Then he returned to Ulster:* This is
episode (g) of the first interpolation omitted from chapter VIII —
see notes to page 127. It replaces a shorter version that occurs
here.

215 *Rochad's Bloodless Fight:* The title may also be read: 'Rochad's
Woman-Fight.'

. . . slaughtering each other there in Glenn Domain: The following
sentence is omitted: 'These are "Rochad's Bloodless Fight" and "the
Mutual Slaughter at Glenn Domain".'

216 *stones and clods:* 'stones' only in the text.

. . . and fell over the camps of Fergus and Ailill: The words 'and
Nera' are omitted here. Nera is a character in one of the *remscéla*
not used for this translation.

216/8 *The Combat of Munremur and Cúroi:* This is episode (c) of the
first interpolation omitted from chapter VIII. It is given here instead
of the following version of the episode:
    This is Amargin's vision. After this came Amargin's vision in
    the land of Tailtiu. He started up out of his dream and none
    dared show his face against him in Tailtiu. It is there that
    Cúroi mac Dáiri came with his army to fight Cúchulainn.
    He heard that Cúchulainn had taken on the men of Ireland

single-handed for the three months of winter. Cúroi mac
Dáiri thought it dishonourable to attack a worn and wounded
man with an army; for Cúchulainn's joints were going from
him. So Cúroi and Amargin started pelting each other and
the stones met in the air. Then Cúroi asked Amargin to leave
off the Táin at Tailtiu. Amargin did so, and Cúroi promised
in return not to march with the armies from then on. It was
done accordingly. But Cúroi started following the armies and
when Amargin saw this he turned his left board toward
Tailtiu and Ráith Airthir and began casting at them once
more. This was the third uncountable thing on the Táin, the
number of his dead. His son Conall Cernach stayed and
supplied him with stones and javelins.

218 *Mag Clochair:* This place is much further north than Tailtiu —
near Focherd. The anomaly is due to the switching of versions, as
noted above.

...*until the day of the Battle:* The following sentence is omitted:
'And Cúroi didn't come until the combat with Ferdia.'

Sualdam's Repeated Warning: See the incident of Sualdam's setting-
out on page 68.

...*how they were harassing his son Cúchulainn:* The text reads:
'how his son Cúchulainn was being harassed by the twelve sons of
Gaile Dána and his sister's son', referring to the episode immediately
before the combat with Ferdia.

219 ...*no man spoke before Conchobor:* reading with the Book of
Leinster text.

...*and the scalloped rim cut his head off:* The following is omitted:
'Others say he was asleep against a stone, and fell on his shield
when he woke.'

220 ...*summon Deda*...*and Leamain and Fallach*...etc.: This list is
not to be taken too seriously, as the inclusion of Cúchulainn and
Cethern indicates. The translation includes some speculative readings as well as details adopted from the Book of Leinster.

221 *Finnchad found that his task was easy*...: The title 'The Vision of
Cormac Connlongas' occurs here in the text, out of place, and is
omitted.

222 *These eight score warriors:* 'They' in the text.

223 *Ailill's hours!*...etc.: The place-names in this *rosc* occur in the
story (except for 'Tuath Bressi' which is taken from the Book of
Leinster text) but are not associated with these incidents.

224 XIII THE COMPANIES ADVANCE.

...from the Sliab Fuait road...: In the sequence of episodes, Sliab Fuait has been left far behind.

225 *the tower of Breogan:* It was from this tower, in Spain, that the Milesian invaders, the Celts, saw Ireland for the first time.

230 *Rochad mac Faithemain from Rígdonn:* 'from Brí Dumae' in the text. The change is made for consistency.

235 *Conchobor will throw up three mounds of men...:* 'Cúchulainn' in the text.

Cúchulainn, wounded in the unequal struggle, hasn't come: A passage is omitted here, beginning 'unless that warrior who came with a single chariot was he...' and continuing with yet another description of Cúchulainn, his horses, his chariot and his charioteer. Cúchulainn arrives only late in the battle.

238 XIV THE LAST BATTLE.

Conchobor came with his armies...: This paragraph is slightly rearranged in translation.

Nét's wives, Nemain and the Badb: The text reads 'Badb and Nét's wife and Nemain.' See note to page 68.

238/9 *That same night... that survived the former slaughter by Cúchulainn:* In the text this passage comes after the first stage of the battle, after ' "Necessity is a great spur," Cúchulainn said' (page 243). A sentence immediately preceding, 'No more is told here about the men of Ulster,' is omitted.

239 *and three named Aed from Aidne:* A further seventy names, approximately, are omitted here.

West of Ferdia's Ford...: Literally, 'To the left of...'

241 *Now some of our servants from the eastern camp are coming out to take them from them:* This sentence is supplied from the Book of Leinster version.

245 *...like a king's horses churning up the ground:* Conjectural. The speech is broken at this point in the text by the words: 'Then Fergus swore this oath.'

246 *...their arms on their elbows, with elbows on wrists...:* The word 'rigthi' is translated as both 'arms' and 'wrists.'

247 *'...on this glutted field of battle,' Fergus said to Ailill:* The following is omitted here:

> Then the Badb and Nemain, Nét's wife, cried out to them that night at Gáirech and Irgairech, and one hundred of their

warriors died of terror. That was not the easiest of nights for them.

247 *A hundred Ulster warriors died by his sword ... a whore's back-side:* This incident occurs later in the text, after 'So Fergus turned aside ...' (page 248), where it is repetitious, and causes some confusion. The following is omitted:
'Soldiers, what can I do?' he said. 'Strike crosswise at the hills, and the warriors around them,' Conall Cernach said.

250 *Medb had set up a shelter of shields ... He wouldn't strike her from behind:* This incident is supplied from the Book of Leinster.
Medb's Foul Place: Literally, 'Medb's Urine'.

251 *... on the day after the battle:* Literally, 'on the morning of the battle.'

The men of Ireland asked who should judge ... between the bulls to judge them: This passage is supplied from the Book of Leinster, in place of the following:
Bricriu the Poison-Tongued had lain sick, in the west, after Fergus had broken his head with a *fidchell* piece.
The reference is to an incident in the tale *Echtra Nerai* (The Adventures of Nera), sometimes counted among the *Táin's* pre-paratory tales; the incident is summarised in the Book of Leinster version as follows:
A year before the time of this story of the Táin Bó Cuailnge Bricriu had travelled from one province to the other to ask something of Fergus. Fergus kept him at his beck and call, waiting for the gifts and valuables in question, and as they were playing *fidchell* a quarrel sprang up between them. Bricriu spoke sharply to Fergus and Fergus struck him with his fist. He had a *fidchell* piece in his hand and drove the piece into Bricriu's head and broke a skull-bone. So all the time the men of Ireland were raiding on the Táin, Bricriu lay healing in Cruachan. The day they returned from the Táin was the day Bricriu got up.
This is, of course, inconsistent with Bricriu's presence on the Táin already mentioned on a number of occasions.

252 *Then the bulls fought each other ... hanging from his horns:* Supplied from the Book of Leinster.

253 *Finnabair stayed with Cúchulainn ...:* This is merely part of the story-teller's final flourish. It ignores Finnabair's death, in the episode 'Rochad's Bloodless Fight' (page 215), and the fact that Emer was Cúchulainn's wife.

Finit, amen: The Book of Leinster version ends with the following scribal notes:

(in Irish) 'A blessing on everyone who will memorise the Táin faithfully in this form, and not put any other form on it,' and (in Latin) 'I who have copied down this story, or more accurately fantasy, do not credit the details of the story, or fantasy. Some things in it are devilish lies, and some poetical figments; some seem possible and others not; some are for the enjoyment of idiots.'